Centre for Innovation in Mathematics Teaching
University of Exeter

STATISTICS

Writers David Cassell
Ian Hardwick
Mary Rouncefield
David Burghes

Editor David Burghes

Assistant Editors Ann Ault
Nigel Price

Heinemann Educational

Heinemann Educational
a division of Heinemann Educational Books Ltd.
Halley Court, Jordan Hill, Oxford OX2 8EJ

OXFORD LONDON EDINBURGH
MADRID ATHENS BOLOGNA PARIS
MELBOURNE SYDNEY AUCKLAND SINGAPORE
TOKYO IBADAN NAIROBI HARARE
GABORONE PORTSMOUTH NH (USA)

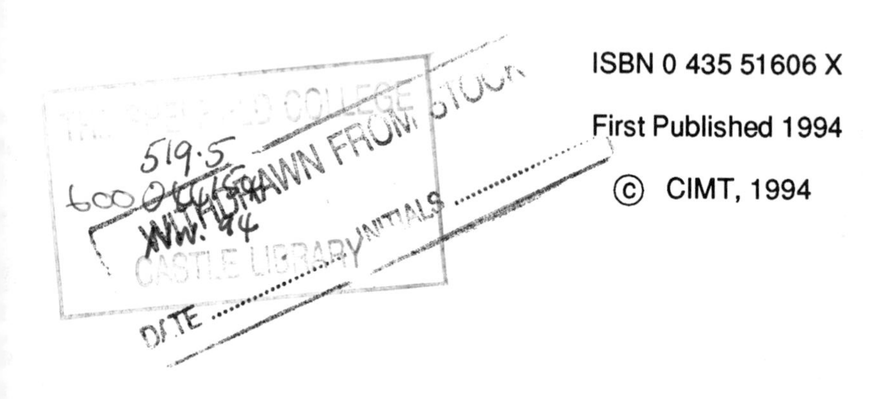

ISBN 0 435 51606 X

First Published 1994

Typseset by ISCA Press, CIMT, University of Exeter

Printed in Great Britain by The Bath Press, Avon

STATISTICS

This is one of the texts which has been written to support the AEB Mathematics syllabus for A and AS level awards first available in Summer 1996.

The text, which covers the Statistics Module, is a development of the Statistics Core text written for the AEB Coursework syllabus.

The development of these texts has been coordinated at the

Centre for Innovation in Mathematics Teaching

at Exeter University in association with Heinemann and AEB.

The overall development of these texts has been directed by David Burghes and coordinated by Nigel Price.

Enquiries regarding this project and further details of the work of the Centre should be addressed to

Margaret Roddick
CIMT
School of Education
University of Exeter
Heavitree Road
EXETER EX1 2LU.

CONTENTS

ACKNOWLEDGEMENTS

PREFACE

Chapter 1 PROBABILITY

1.0	Introduction	1
1.1	Theoretical probability: symmetry	1
1.2	Empirical probability: experiment	5
1.3	Empirical probability:observation	7
1.4	Combined events	9
1.5	Tree diagrams	13
1.6	Conditional probability	16
1.7	Independence	20
1.8	Miscellaneous Exercises	24

Chapter 2 DATA COLLECTION

2.0	Introduction	29
2.1	What sort of data?	31
2.2	Sources of data	33
2.3	Sampling: factors and bias	35
2.4	Miscellaneous Exercises	45

Chapter 3 DESCRIPTIVE STATISTICS

3.0	Introduction	47
3.1	Sorting and grouping	47
3.2	Illustrating data - bar charts	51
3.3	Illustrating data - pie charts	56
3.4	Illustrating data - line graphs and scattergrams	57
3.5	Using computer software	58
3.6	What is typical?	60
3.7	Grouped data	62
3.8	Interpreting the mean	65
3.9	Using your calculator	67
3.10	How spread out are the data?	68
3.11	Standard deviation	73
3.12	Miscellaneous Exercises	80

Chapter 4 DISCRETE PROBABILITY DISTRIBUTIONS

4.0	Introduction	87
4.1	Expectation	88
4.2	Variance	90
4.3	Probability density functions	92
4.4	The uniform distribution	95
4.5	Miscellaneous Exercises	97

Chapter 5 BINOMIAL DISTRIBUTION

5.0	Introduction	99
5.1	Finding the distribution	99
5.2	The mean and variance of the binomial distribution	107
5.3	Miscellaneous Exercises	112

Chapter 6 POISSON DISTRIBUTIONS

6.0	Introduction	115
6.1	Developing the distribution	116
6.2	Combining Poisson variables	121
6.3	The Poisson distribution as an approximation to the binomial	126
6.4	Miscellaneous Exercises	129

Chapter 7 CONTINUOUS PROBABILITY

7.0	Introduction	131
7.1	Looking at the data	132
7.2	Finding a function	134
7.3	Calculating probabilities	138
7.4	Mean and variance	140
7.5	Modes, medians, quartiles	143
7.6	Rectangular distribution	146
7.7	Miscellaneous Exercises	148

Chapter 8 THE NORMAL DISTRIBUTION

8.0	Introduction	151
8.1	Looking at your data	152
8.2	The p.d.f. of the normal	155
8.3	Transformation of normal p.d.f.s	157

8.4 More complicated examples 160
8.5 Using the normal as an approximation to other distributions 162
8.6 A very important application of the normal 169
8.7 Miscellaneous Exercises 170

Chapter 9 ESTIMATION

9.0 Introduction 173
9.1 Sampling methods 174
9.2 Sample size 176
9.3 The distribution of $\overline{X}$ 177
9.4 Identifying unusual samples 179
9.5 Confidence intervals 181
9.6 Miscellaneous Exercises 185

Chapter 10 HYPOTHESIS TESTING

10.0 Introduction 189
10.1 Forming a hypothesis 190
10.2 The sign test 192
10.3 Hypothesis testing for a mean 195
10.4 Hypothesis testing summary 199
10.5 Miscellaneous Exercises 201

Chapter 11 CHI-SQUARED

11.0 Introduction 203
11.1 The Chi-squared table 204
11.2 Contingency tables 208
11.3 Miscellaneous Exercises 212

Chapter 12 CORRELATION AND REGRESSION

12.0 Introduction 215
12.1 Ideas for data collection 215
12.2 Studying results 216
12.3 Pearson's product moment correlation coefficient 219
12.4 Spearman's rank correlation coefficient 224
12.5 Linear regression 230
12.6 Bivariate distributions 234
12.7 Miscellaneous Exercises 237

ANSWERS 243

APPENDIX 251

INDEX 267

PREFACE

Statistical techniques are now used throughout modern industrial societies for analysis of data, to explain observations, and to help make rational decisions. Modern communication techniques have also meant a widespread use of statistical ideas and techniques in presenting and analysing data for public consumption. For example, the way election night results are presented has been revolutionised over the past few decades with all sorts of technical props at the fingertips of the presenters. But it should be remembered that statistics alone does not solve problems and indeed can come up with the wrong conclusions. This was borne out by the failure of political pollsters to correctly predict the result of the 1992 election in the U.K.

Nevertheless this is just one instance of where statistical analysis did not provide correct answers; there are many instances when it does, and perhaps more importantly, instances where statistical analysis helps decision making. We live in a stochastic world, that is, one that is not predetermined. Governments and local authorities use statistical techniques for planning, developments, allocating resources, controlling the economy and indeed gauging public opinion on topical isues. The range and scope of the techniques available has increased considerably over the past few decades, and most importantly the technology is now available to cope, relatively cheaply, with large amounts of data. Even the smallest business can afford its own computing power which can be used to monitor and predict future trading patterns.

This text has been written with the aim of giving readers a thorough understanding of statistical ideas and concepts, based on **probability theory**. It is not a recipe of what to do - there are plenty of good texts that fit that bill already - but attempts to show why a particular technique is used as well as explaining how to use it. We want readers to get a feel for statistics - both its potential and its limitations. There are many worthy techniques not included in this text, but readers gaining a sound understanding of probability and statistics should have little difficulty in coping with these techniques if they are needed later.

This text has been produced for students and includes examples, activities and exercises. It should be noted that the activities are **not** optional but are an important part of the learning philosophy in which you are expected to take a very active part. The text integrates

- **Exposition** in which the concept is explained;
- **Examples** which show how the techniques are used;
- **Activities** which either introduce new concepts or reinforce techniques;
- **Discussion Points** which are essentially 'stop and think' points, where discussion with other students and teachers will be helpful;
- **Exercises** at the end of most sections in order to provide further practice;
- **Miscellaneous Exercises** at the end of each chapter which provide opportunities for reinforcement of the main points of the chapter.

Discussion points are written in a special typeface as illustrated here.

Note that answers to the exercises are given at the back of the book. You are expected to have a calculator available throughout your study of this text and occasionally to have access to a computer.

Some of the sections, exercises and questions are marked with an asterisk (*). This means that they are either **not** central to the development of the topics in this text and can be omitted without causing problems, or they are regarded as particularly challenging.

This text is one of a series of texts written specially for the new AEB Mathematics syllabus for A and AS level coursework. The framework is shown opposite. Essentially each module corresponds to an AS level syllabus and two suitable modules provide the syllabus for an A level award. Optional coursework is available for students taking any of the three applied modules Mechanics, Statistics and Discrete Mathematics.

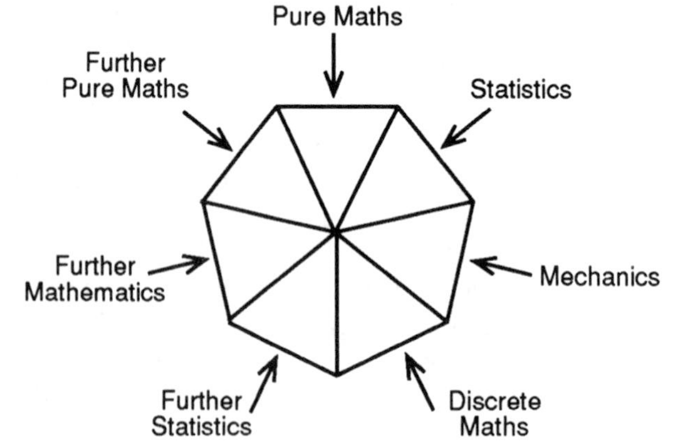

Full details of the scheme are available from
AEB, Stag Hill House, Guildford GU2 5XJ.

We hope that you enjoy working through the book. We would be very grateful for comments, criticisms and notification of any errors. These should be sent to

Margaret Roddick
CIMT
School of Education
University of Exeter
EXETER EX1 2LU.

ACKNOWLEDGEMENTS

This text has been developed from an earlier version, published specifically for the CIMT/AEB Mathematics syllabus and based on the philosophy of the Wessex Project. I am particularly grateful to the original authors and to Bob Rainbow (Wessex) and John Commerford (AEB) for their help and encouragement in the development of the original resources.

This revised version and associated texts have been written for the new AEB Mathematics syllabus and assessment, which will be examined for the first time in Summer 1996. I am grateful for the continued support from AEB through its mathematics officer, Jackie Bawden, and to the staff at Heinemann, particularly Philip Ellaway.

Finally, I am indebted to the staff at CIMT who work with dedication and good humour despite the pressure which I continually put them under. In particular, I am grateful to Nigel Price for help with editing of the text, to Ann Ault for checking the mathematics and to Liz Holland, Margaret Roddick and Ann Tylisczuk for producing camera ready copy.

David Burghes

(Project Director)

1 PROBABILITY

Objectives

After studying this chapter you should

- understand how the probability of an event happening is measured;
- recognise whether or not events are related in any way;
- be able to assess the likelihood of events occurring.

1.0 Introduction

'Sue is more likely than Jane to be head girl next year.'

'It will probably rain for the fete tomorrow.'

'A European football team has a better chance of winning the next world cup than a South American one.'

'Reza is 'odds on' to beat Leif in the chess final.'

All these sentences express an opinion that one outcome is more likely than another but in none of them is there any attempt to say by how much. Yet if you want to take out insurance against bad weather for the fete the insurance company you approach must have a way of calculating the probability or likelihood of rain to know how much to charge.

So how can you assess the chance that some event will actually happen?

1.1 Theoretical probability: symmetry

Many intuitive ideas of chance and probability are based on the idea of **symmetry**. Consider the following questions:

If you toss a coin repeatedly, how many times will it come down heads?

If you roll a die how often will you get a four?

If you roll two dice several times, how often will you get two sixes?

For the second question, your answer should be about one in six times provided the die is a fair one. Another way of expressing this is to say that the probability of obtaining 4 is

$$\frac{1}{6} \Rightarrow p(4) = \frac{1}{6}.$$

The answer is dependent on the idea of symmetry. That is, every possible outcome (namely 1, 2, 3, 4, 5 and 6) is equally likely to occur. So the probability of any one score must be $\frac{1}{6}$.

Sometimes, though, you must be very careful to make sure that you have a complete list of **all** the possible outcomes of the event under consideration.

Activity 1 The three card game

Suppose you have three cards:

Card A is white on both sides

Card B is black on both sides

Card C is black on side 1 and white on side 2.

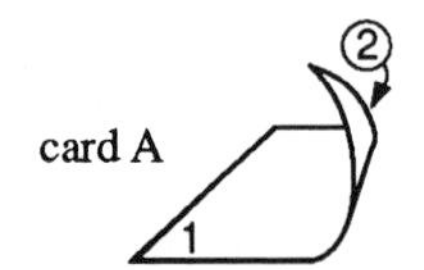

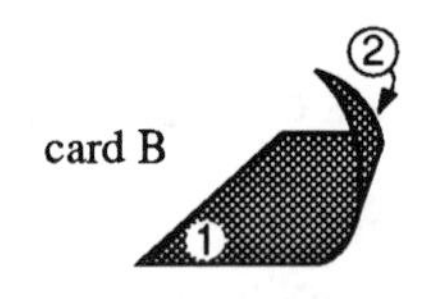

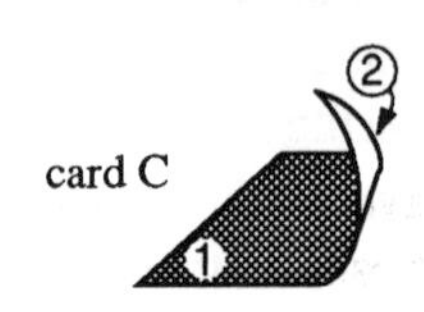

You shuffle them and place them in a pile on the table so that you can see only the upper face of the top card, which is black.

If I were to say,

> "I will pay you £5 if the reverse face of the top card is white and you pay me £3 if it is black."

should you take the bet?

If you said that there are two possibilities - the lower face is either black or white - then this is certainly correct. However, if you have gone on to decide that you are just as likely to win as to lose then perhaps you have not listed all the possible cases.

With the three cards, if you can see a black face then the three possibilities are that you are looking at

side 1 of Card C side 1 of Card B side 2 of Card B

and since two of these (side 1 and side 2 of Card B) have black on the reverse, the bet is not a good one for you.

In the long run you would win £5 once in three games and lose £3 twice so you can expect to lose £1 on average every three games or $33\frac{1}{3}$p per go.

Activity 2

Play the three card game a number of times with a friend. You can use either cards, as shown on the previous page, or a die with 1, 2, 6 painted black and 3, 4, 5 white. Remember to always bet on the same colour being underneath as is showing on top.

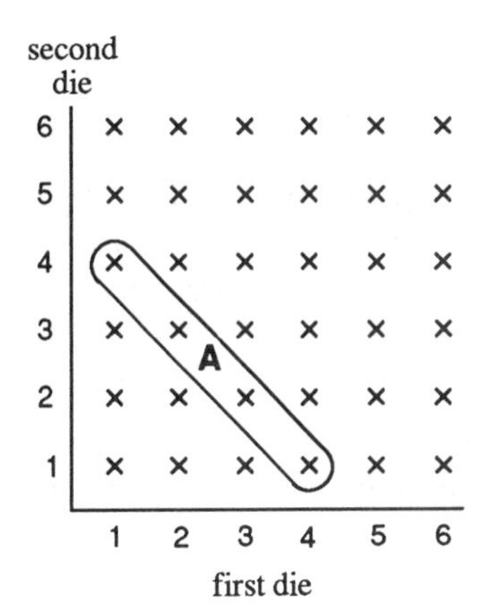

Listing all the equally likely outcomes can be very tedious so you may find it simpler and clearer to show them in a diagram.

For example, when two dice are rolled there are thirty six possible outcomes which can be shown very neatly in a diagram (see opposite).

This is called the **sample space**. You can see by looking at the crosses in the area labelled **A** that, for example,

$$P(\text{total} = 5) = \frac{4}{36} = \frac{1}{9}.$$

This sort of diagram can be adapted to other problems so it is very useful.

What is $P(\text{total} = 7)$?

Example

Two of the five reserves for the school ski trip, Tamsin, John, Atanu, Robin and David can have places now that a couple of people have had to drop out. How likely is it that John and Tamsin will be chosen to go?

Solution

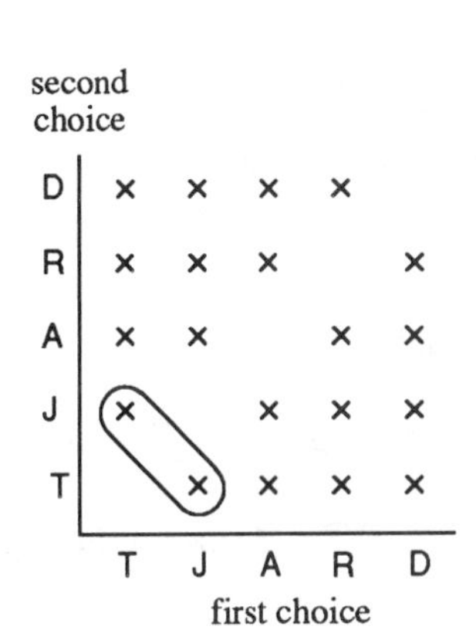

Only the two cases indicated out of the twenty in the diagram opposite are situations when John and Tamsin are chosen, so

$$P(\text{T and J}) = \frac{2}{20} = \frac{1}{10}.$$

You know that with **one** die there are six different possible outcomes and the diagram for **two** dice showed that there are thirty six possible outcomes in this case.

How many will there be if three dice are used?

What sort of diagram could be drawn to show the different results?

As one die needs a one-dimensional diagram which gives six possibilities and two dice need a two-dimensional diagram to show thirty six outcomes, a sensible idea to try for three dice would be a three-dimensional picture.

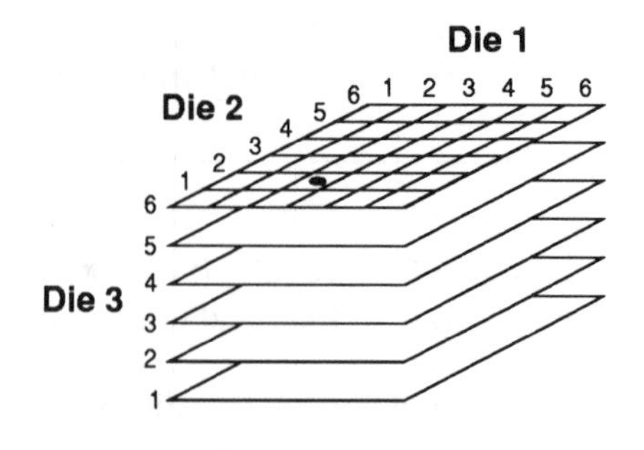

The diagram opposite shows six of the two-dimensional diagrams in layers on top of each other so there are $6 \times 36 = 216$ possibilities in this case or 6^3.

The • in the diagram represents 3 on the first die, 2 on the second and 6 on the third.

The number of dice used appears as a power in these examples so it should be possible to work out the total number of outcomes when more than three dice are used.

Example

What is the probability of getting five sixes when five dice are rolled?

Solution

Five dice produce $6^5 = 7776$ outcomes.

Only one outcome is all sixes, so

$$P(\text{five sixes}) = \frac{1}{7776}.$$

Example

What is the probability that there will be at least one head in five tosses of a fair coin?

Solution

Five coins produce $2^5 = 32$ outcomes.

Only T T T T T does not contain at least one head, so

$$P(\text{at least one H}) = \frac{31}{32}.$$

Exercise 1A

1. What is the probability of choosing an even number from the set of numbers {1, 2, 3, 5, 6, 7, 8, 10}?
2. When two six-sided dice are rolled what is the probability that the product of their scores will be greater than six?
3. If you have three 10p coins and two 50p coins in your pocket and you take out two at random, what is the probability that they add up to 60p?

   (Draw a sample space.)
4. If two people are chosen at random what is the probability that they were born on the same day of the week?
5. List the ways in which one head and five tails may be obtained from six tosses of a coin.
6. Two dice are rolled and the 'score' is the product of the two numbers showing uppermost. If the probability is $\frac{11}{36}$ that the score is at least N, what is the value of N?
7. Pierre and Julian each roll one die. If Pierre's shows the higher number then he wins 7p, otherwise he loses 5p. Explain why this is fair. If Pierre were to add three dots to convert the two on his die to a five, how will it affect his winning?
8. A card is chosen at random from a pack of fifty-two. It is replaced and a second card is selected. What is the probability that at least one is a picture card (Jack, Queen, King)? (Sketch a sample space but don't bother with all the crosses.)
9. Eight people are seated at random round a table. What is the probability that Sharif and Raijit will be next to each other?

1.2 Empirical probability: experiment

Mathematicians' early interest in the subject of probability in the seventeenth century came largely as a result of questions from gamblers in France. Since dice, cards, etc. were used, the situations involved had outcomes which were equally likely. All the arguments then could be based on symmetry. You must also be prepared, however, for other situations which do not have properties of symmetry.

It was possible to answer the question about the die as there were six possible outcomes which were equally likely to occur as the cube is a simple regular solid. However, you might find questions about a **cuboctahedron** not as simple to answer.

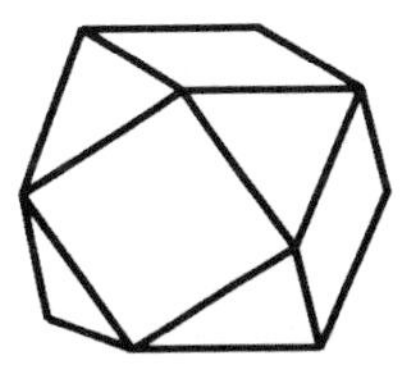

Cuboctahedron

This solid is formed by cutting equilateral triangles from the corners of a cube to produce six square and eight triangular faces.

What is the probability of the solid ending with a square facing upwards when it is rolled?

Perhaps it depends on how many of the faces are squares. Or does considering the areas of the squares as a fraction of the total surface area seem more likely?

Without testing and evidence nobody will believe any answer you give to the question so you will need to experiment to find the probability of a square facing upwards.

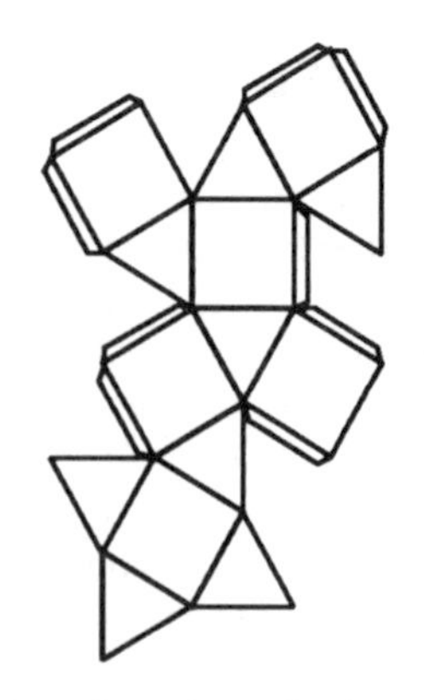

Net of a cuboctahedron

Activity 3

Find the answer for yourself by making the solid from a copy of the net. Be prepared to roll it many times.

You can see the probability graphically by plotting the number of rolls on the x-axis and the fraction of the times a square is facing upwards on the y-axis.

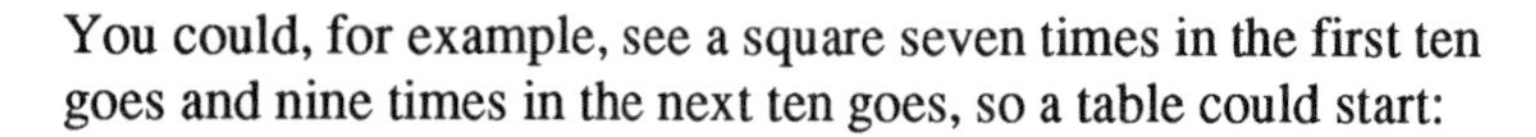

You could, for example, see a square seven times in the first ten goes and nine times in the next ten goes, so a table could start:

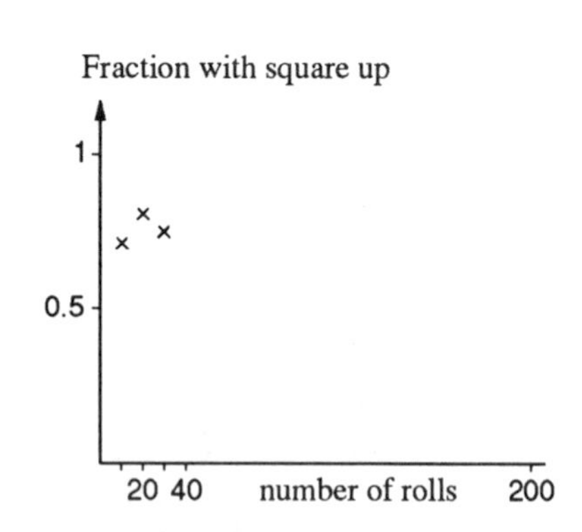

No. of squares in 10 goes	7	9	
Total no. of squares	7	16	
Total no. of rolls	10	20	30
Fraction (probability)	$\frac{7}{10} = 0.7$	$\frac{16}{20} = 0.8$	

Of course, if several people are doing this experiment you could put your results together to achieve a more reliable answer.

What you will often find from experiment is that the fraction calculated will gradually cease to vary much and will become closer to the value called the **probability.**

What is the probability of a square not appearing uppermost?

From this experiment you will have produced an **empirical probability**; i.e. one based on experience rather than on a logical argument.

The idea of experiment and observation then gives a probability equal to

$$\frac{\text{the number of times a square was upwards}}{\text{the number of attempts}}$$

So if you saw it happen on 150 out of 200 times you will have come to the conclusion that

$$P(\text{square}) = \frac{150}{200} = \frac{3}{4}.$$

In reality, if the true probability was $\frac{3}{4}$, you would be unlikely to get exactly 150 out of 200 – but you should be somewhere near it.

Activity 4 Coin tossing

Toss an unbiased coin 100 times, and record the total fraction of heads after every 10 goes. Plot these on a graph of fraction of heads against number of goes. Does this indicate that the coin is a fair one?

1.3 Empirical probability: observation

How likely is it that the writer of this text is alive now? It is hard to conduct an experiment on this but if I am forty now and writing this in 1991 then you can make use of observations on the life expectancy of forty-year-old males.

Male age	Average life expectancy beyond present age	Probability of surviving at least 5 years
35	38.1	0.993
36	37.1	
37	36.2	
38	35.2	
39	34.3	
40	33.3	0.988
41	32.4	
42	31.5	
43	30.5	
44	29.6	
45	28.7	0.979

Data like these are needed by insurance companies for their life policies. Some people will look at tables of figures for sunshine hours and rainfall when planning holidays.

Probability is of interest to people working in economics, genetics, astronomy and many other fields where it may be difficult to experiment but where data can be gathered by observation over a long period.

Example

Jane travels to school on the train every weekday and often sees rabbits in a field by the track. In four weeks her observations were

Number of rabbits seen	0	1	2	3	4	5	6	7	8
Number of occasions	0	3	5	7	2	1	0	1	1

What is the probability that on her next journey she will see at least two rabbits?

Solution

$$P(\text{at least two rabbits}) = \frac{17}{20} = 0.85,$$

as on $5+7+2+1+0+1+1 = 17$ days out of the 20 she saw two or more rabbits.

Exercise 1B

1. Using the information from the example above, what is the probability that Jane sees:

   (a) 3 or 4 rabbits; (b) 6 rabbits;
   (c) at least one rabbit?

2. The number of visitors to the UK from North America in 1988 is given below in categories to show mode of travel and purpose of visit.

	Air	Sea
Holiday	1269	336
Business	605	17
Friends and relatives	627	55
Miscellaneous	324	39

(in 1000s)

   If you were to have met a visitor from North America in 1988 what would have been the probability that the visitor

   (a) was here on business;

   (b) came by sea;

   (c) came by air to visit friends or relatives;

   (d) was here on business if you know the visitor came by sea?

3. Draw a circle of radius 5 cm and add a square of side 10 cm so that the circle touches its four sides.

   Take random numbers from a table, four at a time, and interpret them as co-ordinates using the bottom left hand corner of the square as the origin. (For example, the numbers 4 6 2 0 give the point (4.6, 2.0) with measurements in cm.)

   Use a large number of points and see what fraction of them lie inside the circle.

   (The area of the square is 100 units and the area of the circle is $\pi 5^2 = 25\pi$. The fraction of the square taken up by the circle is

$$\frac{25\pi}{100} = \frac{\pi}{4}$$

   so your result is an approximation to $\frac{\pi}{4}$ and can be used to estimate π.)

4. Take ten drawing pins and drop them onto a flat surface. Note how many finish point up. Repeat this several times and produce a table and graph like those you used with the cuboctahedron.

   What is the probability that a drawing pin accidentally dropped will fall into a point-up position?

1.4 Combined events

Complement

In the probability experiment in Section 1.2 you will have obtained a value for probability by considering, for example, the number of times a square face finished uppermost as a fraction of the total number of rolls as

$$P(\text{square}) = \frac{\text{no. of times square finished upwards}}{\text{no. of trials}}$$

This is also called **relative frequency**.

The largest value this fraction can take is one, when a square face appears every trial, and the smallest it can be is zero, when a triangle is uppermost on each go, so

$$0 \leq \text{probability} \leq 1.$$

Another result that may be obvious is that the number of times with a square facing up plus the number of times with a triangle facing up equals the number of trials.

Hence

$$\frac{\text{no. of times with square up}}{\text{no. of trials}} + \frac{\text{no. of times triangle up}}{\text{no. of trials}} = 1$$

$$\Rightarrow \quad P(\text{square}) + P(\text{not square}) = 1$$

which is written in general as

$$\boxed{P(A) + P(A') = 1}$$

where A' means 'not A' or the '**complement** of A'.

You may well have used this idea earlier when you answered the question in Section 1.2 about how likely it is for a square not to appear on the top face when a cuboctahedron is rolled.

Intersection

Take a cube and mark on its different faces three black circles, one black cross and two red crosses.

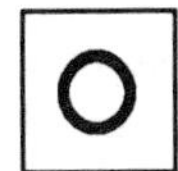
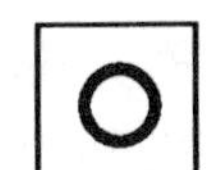
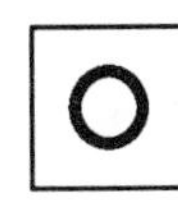
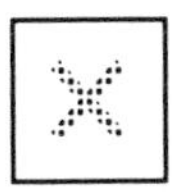
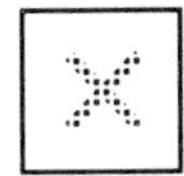

When it is rolled, what are the probabilities of getting

$P(\text{red})$, $P(\text{black})$, $P(\text{circle})$ **and** $P(\text{cross})$ **?**

What is the likelihood of getting a black symbol and a cross?

You can see that just one of the cube's six faces is covered by this description, so

$$P(\text{black and cross}) = \frac{1}{6}.$$

This can be written as

$$P(\text{black} \cap \text{cross}) = \frac{1}{6}.$$

Another way of showing all the possibilities is illustrated opposite.

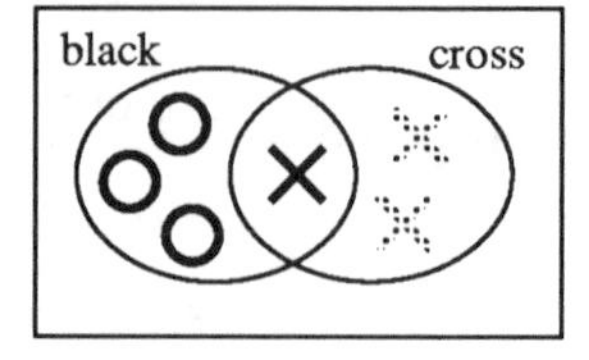

or

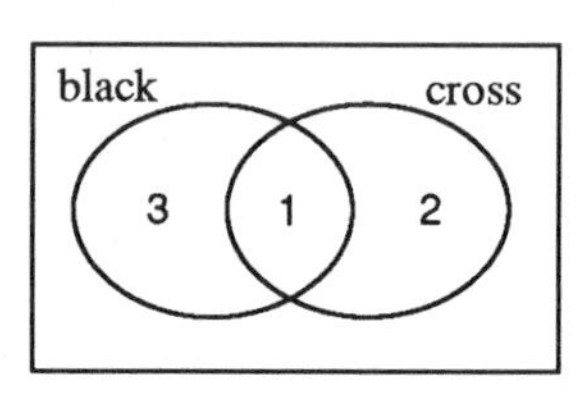

These are called **Venn diagrams** after *John Venn*, an English mathematician and churchman, who studied logic and taught at Cambridge.

What is the value of $P(\text{red} \cap \text{cross})$?

You may have noticed that

$$P(\text{red} \cap \text{cross}) + P(\text{black} \cap \text{cross}) = P(\text{cross}).$$

If you were asked for the probability of a circle **and** a red symbol finishing uppermost from a single roll you should realise that

$$P(\text{red} \cap \text{cross}) = 0$$

as the two cannot happen at the same time. These are called **mutually exclusive** events as the occurrence of either excludes the possibility of the other one happening too.

Union

Eight teams are entered for a knock-out netball tournament and two of these are the YWCA and the Zodiac youth club.

What is the probability that the YWCA or Zodiac will reach the final?

('or' here means one or the other or both, more technically called the inclusive disjunction.)

How the competition will run is shown opposite but until the draw is made no names can be entered.

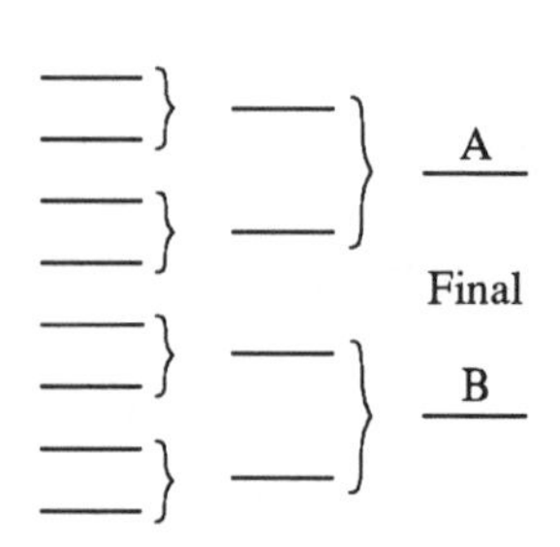

A diagram like the one you used earlier shows all the different possible ways in which the two final places A and B may be filled by the competing teams.

From the figure opposite you can see that

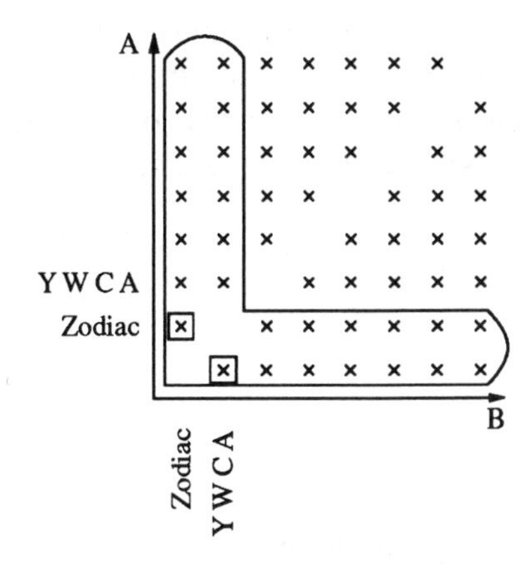

(a) $P(\text{Zodiac in final}) = \dfrac{14}{56}$

(b) $P(\text{YWCA in final}) = \dfrac{14}{56}$

(c) $P(\text{Zodiac or YWCA}) = \dfrac{26}{56}.$

Note that $P(\text{Zodiac or YWCA}) \neq P(\text{Zodiac}) + P(\text{YWCA})$.

Why would you expect these not to be equal?

When the first two probabilities (a) and (b) were worked out, the two cases marked with squares in the diagram were included in each answer. When the probabilities are added together, these probabilities have been counted twice.

These correspond to the two ways of having both Zodiac and YWCA in the final. Their probability is given by

$$P(Z \cap Y) = \frac{2}{56}$$

and you can see that

$$P(Z \text{ or } Y) = P(Z) + P(Y) - P(Z \cap Y).$$

Taking off the $P(Z \cap Y)$ ensures that these two events are not counted twice.

Checking with the figures you get

$$\frac{14}{56} + \frac{14}{56} - \frac{2}{56} = \frac{26}{56}$$

which is true.

Now if you look back to the die marked with circles and crosses you will see that

$$P(\text{black}) = \frac{2}{3}, \quad P(\text{circle}) = \frac{1}{2}$$

so that if you tried to say that

$$P(\text{black or circle}) = P(\text{black}) + P(\text{circle})$$

you would get $P(B \cup C) = \frac{1}{2} + \frac{2}{3} = 1\frac{1}{6}$, where $B \cup C$ means B

or C. This looks decidedly dubious as you know that probability is measured on a scale from zero to one! The problem is that once more you have counted some of the possibilities twice as they are in both categories. Again, if you try

$$P(B \cup C) = P(B) + P(C) - P(B \cap C)$$

then a true statement results:

$$\frac{2}{3} + \frac{1}{2} - \frac{1}{2} = \frac{2}{3}.$$

The $\frac{2}{3}$ on the left is correct as four of the six faces have a black colour or a circle or both.

Is it ever true that $P(A \cup B) = P(A) + P(B)$?

If it is, then $P(A \cap B)$ must be zero and this means that the events are what is called **mutually exclusive**. A Venn diagram could be drawn and would look like the one here with no overlap. So if

$$P(A \cap B) = 0$$

then $$P(A \cup B) = P(A) + P(B).$$

In general though,

$$\boxed{P(A \cup B) = P(A) + P(B) - P(A \cap B)}$$

and this can be illustrated by the Venn diagram opposite. The **intersection** of the two sets, $A \cap B$, is shown whilst the **union**, $A \cup B$, is given by everything inside A and B.

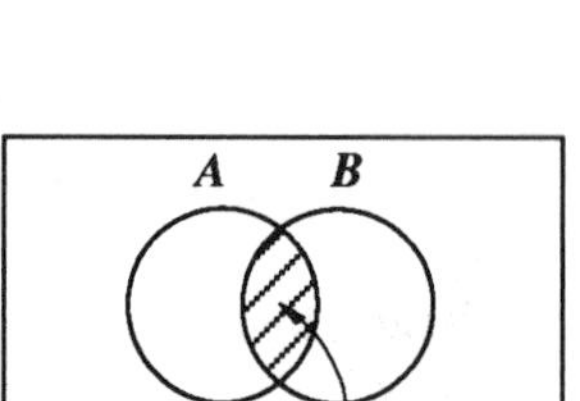

Exhaustive probabilities

The cube you looked at marked with crosses and circles had faces as shown opposite.

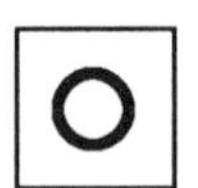
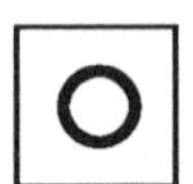
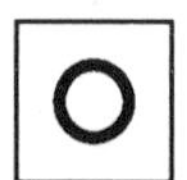
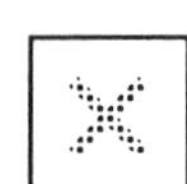

What is the value of $P(\text{black} \cup \text{cross})$?

Since each of six symbols was black or a cross then

$$P(\text{black} \cup \text{cross}) = 1$$

and the events 'getting a black symbol 'and 'getting a cross' are said to form a pair of **exhaustive events**. Between them they exhaust all the possible outcomes and therefore all the probability, i.e. one.

So, if A and B are exhaustive events

$$\boxed{P(A \cup B) = 1}$$

Exercise 1C

1. In a class at school $\frac{1}{2}$ of the pupils represent the school at a winter sport, $\frac{1}{3}$represent the school at a summer sport and $\frac{1}{10}$ do both. If a pupil is chosen at random from this group what is the probability that someone who represents the school at sport will be selected?
2. If the probability that Andrea will receive the maths prize this year is $\frac{1}{3}$ and the probability that Philson will win it is $\frac{1}{4}$,what is the chance that one of them will get it?
3. In a certain road $\frac{1}{5}$ of the houses have no newspapers delivered. If $\frac{1}{2}$ have a national paper and $\frac{2}{3}$ have a local paper, what is the probability that a house chosen at random has both?
4. Consider the following possible events when two dice, one red and one green, are rolled:

   A : the total is 3

   B : the red is a multiple of 2

   C : the total is ≤ 9

   D : the red is a multiple of 3

   E : the total is ≥ 11

   F : the total is ≥ 10.

   Which of the following pairs are exhaustive or mutually exclusive?

   (a) A, B (b) A, D (c) C, E

   (d) C, F (e) B, D (f) A, E

1.5 Tree diagrams

Another approach to some of the problems examined earlier would be to use **'tree diagrams'**. These are sometimes called decision trees and may be used in other subjects such as business studies.

Example

While on holiday, staying with Rachel in Kent in the South East of England, Gabrielle saw a very large black bird. Rachel noticed that it was, in fact, not **all** black and they looked in a bird book to find what it might have been. The facts they discovered are shown in the tree diagram opposite.

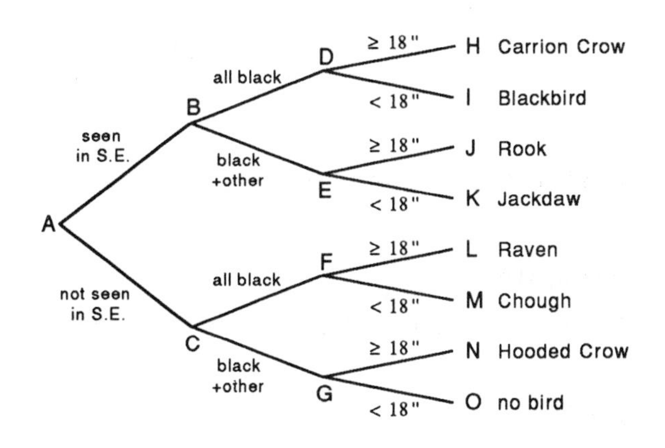

By following along the branches from the left to the right can you decide what they actually saw?

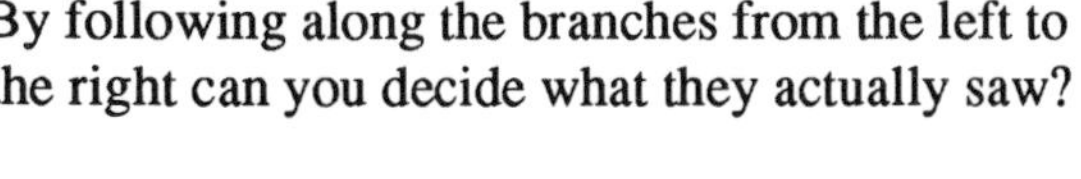

Solution

As they were in Kent you should have moved from A to B. Since the bird was not all black, B to E is the correct choice next, and if the bird was very large then E to J tells you it was a rook.

When you see people flocking from college to the local pub at lunchtime you might be able to identify the individuals by using the tree diagram opposite.

Now a shabbily dressed beer drinking worried person is a teacher. A happy, shabbily dressed beer drinker is a student, so keep smiling!

well dressed – beer – worried – Head
well dressed – beer – happy – Lab. Assistant
well dressed – spirit – worried – Deputy Head
well dressed – spirit – happy – Secretary
shabby – beer – worried – Teacher
shabby – beer – happy – Student
shabby – spirit – worried – Head of Dept.
shabby – spirit – happy – Caretaker

As a result of observation over a long period you might notice that 80% of those who come in are shabby. 90% of these and one third of the others are seen to drink beer. Three quarters of beer drinkers and half of those who prefer spirits look happy. If you put these proportions on the branches as fractions you are in a position to work out how those who come in are divided up as students, teachers, etc.

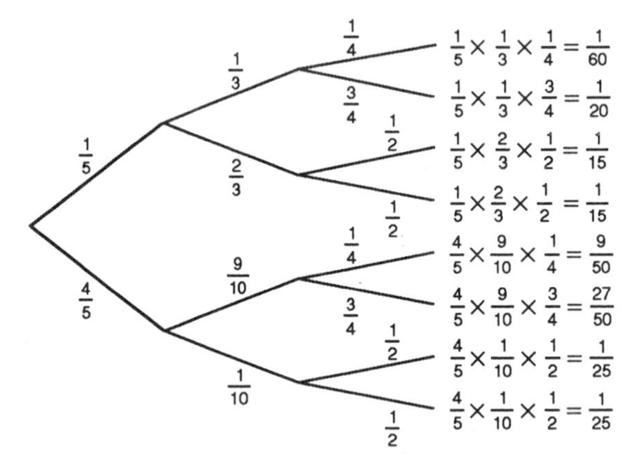

Example

What proportion are teachers?

Solution

The 'teacher' branch is

shabby – beer – worried.

The proportion that are shabbily dressed and drink beer is

$$\frac{9}{10} \text{ of } \frac{4}{5} = \frac{9}{10}\times\frac{4}{5} = \frac{18}{25}.$$

Of those, $\frac{1}{4}$ are worried, giving the proportion of teachers as

$$\frac{1}{4} \text{ of } \frac{18}{25} = \frac{1}{4}\times\frac{18}{25} = \frac{9}{50} \quad (=0.18).$$

What fraction of customers from the college are secretaries?

Example

What fraction of the customers from the college look worried?

Solution

The proportions in each category are shown on the tree diagram.

$$\text{So proportion worried} = \frac{1}{60}+\frac{1}{15}+\frac{9}{50}+\frac{1}{25} = \frac{91}{300} \quad (\approx 0.3).$$

Why is the sum of all the proportions in the tree diagram on the previous page one?

Example

If there are equal numbers of boys and girls in your school and you know that

$\frac{1}{4}$ of the boys and $\frac{1}{10}$ of the girls walk in every day,

$\frac{1}{3}$ of the boys and $\frac{1}{2}$ of the girls get a lift

and the rest come by coach, determine

(a) the proportion of the school population that are girls who go by coach;

(b) the proportion of the school population that go by coach.

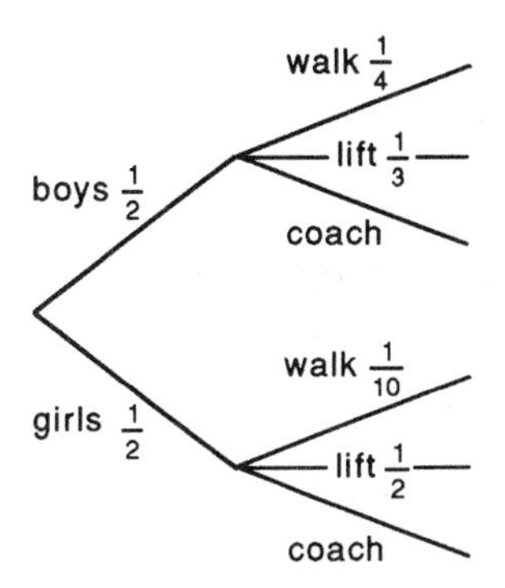

Solution

The branches have missing entries but these can be calculated from the facts already known. Since

$$\frac{1}{4}+\frac{1}{3}=\frac{7}{12}$$

of the boys have been accounted for, there remains $\frac{5}{12}$ who must use the coach.

Similarly, the proportion of girls going by coach is given by

$$1-\left(\frac{1}{10}+\frac{1}{2}\right)=\frac{4}{10}=\frac{2}{5}.$$

All the values are entered on the diagram opposite, so that the answers to (a) and (b) are now easy to see.

(a) $\frac{1}{2}\times\frac{2}{5}=\frac{1}{5}$

(b) $\frac{1}{5}+\frac{1}{2}\times\frac{5}{12}=\frac{49}{120}.$

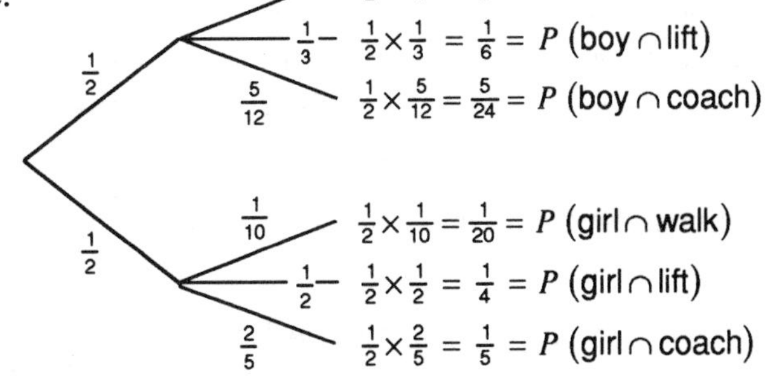

Example

When Sam and Jo play in the hockey team the probability that Sam scores is $\frac{1}{3}$ and that Jo scores is $\frac{1}{2}$, regardless of whether or not Sam does.

What is the probability that neither will score in the next game?

Solution

The tree diagram opposite shows that the answer is $\frac{1}{3}$.

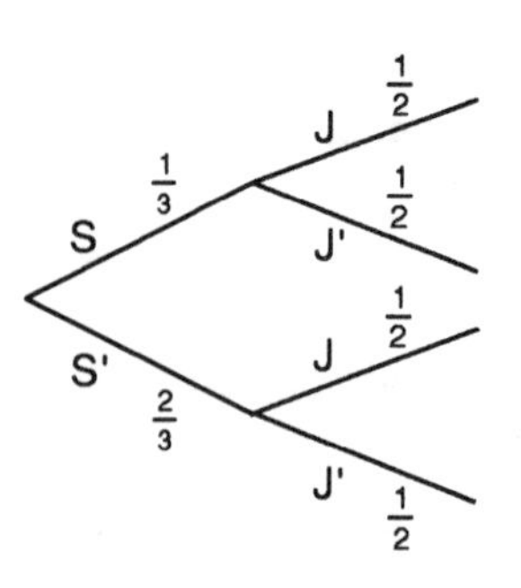

Exercise 1D

1. The probability that a biased die falls showing a six is $\frac{1}{4}$. The biased die is thrown twice.

   (a) Draw a tree diagram to illustrate the probabilities of 'throwing a six' or 'not throwing a six'.

   (b) Find the probability that exactly one six is obtained.

2. In each round of a certain game a player can score 1, 2, 3 only. Copy and complete the table which shows the scores and two of the respective probabilities of these being scored in a single round.

Score	1	2	3
Probability	$\frac{4}{7}$		$\frac{1}{7}$

3. A bag contains 7 black and 3 white marbles. Three marbles are chosen at random and in succession, each marble being replaced after it has been taken out of the bag.

   Draw a tree diagram to show all possible selections.

   From your diagram, or otherwise, calculate, to 2 significant figures, the probability of choosing:

   (a) three black marbles;

   (b) a white marble, a black marble and a white marble in that order;

   (c) two white marbles and a black marble in any order;

   (d) at least one black marble.

   State an event from this experiment which, together with the event described in (d), would be both exhaustive and mutually exclusive.

1.6 Conditional probability

Your assessment of how likely an event is to occur may well depend on some other event or variable. If you were asked, "What is the probability that it will rain next Monday?" your answer would depend on the time of year you were asked. If the question were in winter then $\frac{1}{2}$ might be a realistic assessment but in summer your reply might be $\frac{1}{10}$. This can be written

$$P(\text{rain next Monday} \mid \text{summer}) = \frac{1}{10},$$

that is, the probability of rain next Monday, given that it is summer, is one tenth.

This probability depends on a definitely known condition (that is, it is summer), hence the term '**conditional probability**'.

As another example, consider the following problem:

If the probability of a school pupil wearing glasses is $\frac{1}{9}$, does it make any difference to how likely you think it is that the next one you see will wear glasses if you know that the pupil is female?

Is P (wearing glasses) **the same as** $P(\text{wearing glasses}|\text{female})$**?**

It should be possible to find out by considering a large sample, perhaps when having lunch or at a main entrance.

There is nothing new in the idea of conditional probability and you may well have realised that you have used it already. Conditional probabilities appeared on branches of the tree diagrams to do with the pub's customers and pupils' transport in the last section. The fractions on the branches after the initial ones were conditional probabilities as they definitely depended on the previous ones. The transport tree could have been labelled

$$B = \text{boy},\ G = \text{girl},\ W = \text{walk},\ \ L = \text{lift},\ \ C = \text{coach}.$$

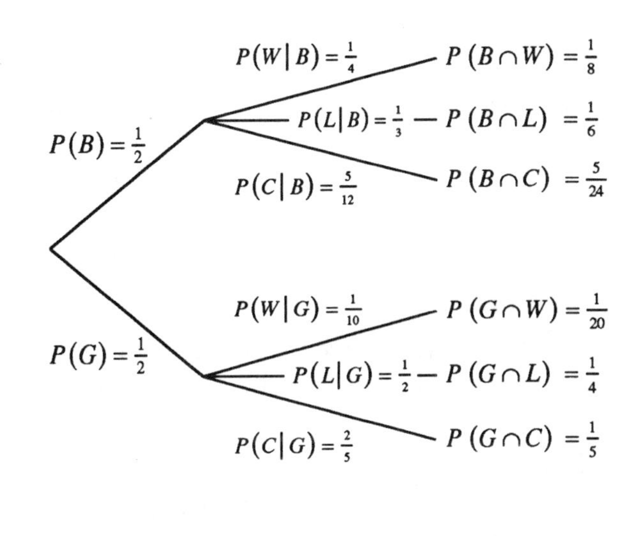

You can readily see that $P(B\cap W)=\frac{1}{8}$ since

$$P(B\cap W) = \frac{1}{2}\times\frac{1}{4} = \frac{1}{8}.$$

Now $\quad P(B)=\frac{1}{2}$, and $P(W|B)=\frac{1}{4}$,

leading to $\quad P(B\cap W) = P(B)\times P(W|B)$

or

$$\boxed{P(W|B) = \frac{P(W\cap B)}{P(B)}}$$

This is a very useful equation when working with conditional probability and holds in general. That is, if A and B are two events,

then

$$\boxed{P(A|B) = \frac{P(A\cap B)}{P(B)}}$$

Example

Using the example from page 14, what is the probability that a worried person from the college is a teacher?

Solution

$$P(\text{teacher} \mid \text{worried}) = \frac{P(\text{teacher} \cap \text{worried})}{P(\text{worried})}$$

$$= \frac{P(\text{teacher})}{P(\text{worried})}$$

$$= \left(\frac{4}{5} \times \frac{9}{10} \times \frac{1}{4}\right) \div \frac{91}{300}$$

$$= \frac{9}{50} \div \frac{91}{300}$$

$$= \frac{9}{50} \times \frac{300}{91}$$

$$= \frac{54}{91}.$$

So now you know what fraction of the worried people are teachers.

Conditional probabilities can also be found from sample space diagrams.

Example

If you roll two dice, one red and one green, what is the probability that the red one shows a six if the total on the two is 9?

Solution

Since you know that the total is 9 you need only look at the four crosses enclosed by the curve in the diagram opposite as they indicate all the possible ways of getting the 9 required. Now just considering these four, what is the chance that the red one shows 6?

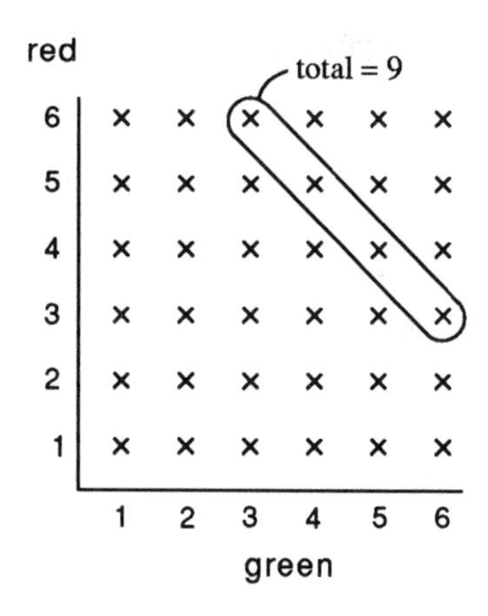

$$P(r = 6 \mid r + g = 9) = \frac{1}{4}$$

as only one of the four crosses has a six on the red.

Example

Class 7C has 18 boys and 12 girls in it and 7K is made up of 12 boys and 16 girls. If you pick one of their registers and a pupil from it at random, what is the probability that you select

(a) a girl (b) from 7C if the choice is a girl?

Solution

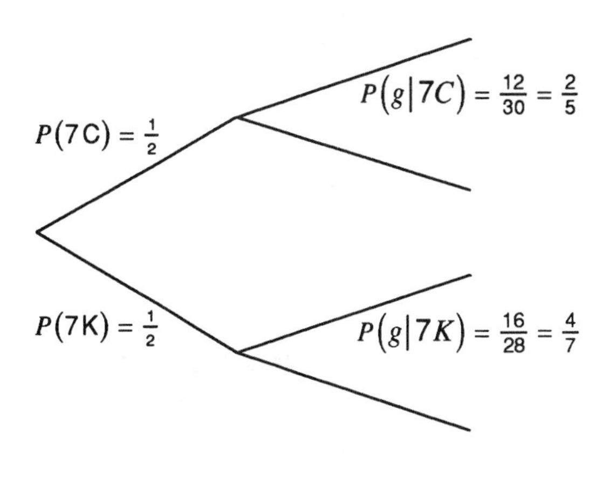

(a) $P(\text{girl}) = \frac{1}{2}\times\frac{2}{5}+\frac{1}{2}\times\frac{4}{7}$

$= \frac{1}{5}+\frac{2}{7} = \frac{17}{35}.$

(b) $P(7C \mid \text{girl}) = \frac{P(7C\cap\text{girl})}{P(\text{girl})}$

$= \frac{\frac{1}{2}\times\frac{2}{5}}{\frac{17}{35}}$

$= \frac{1}{5}\div\frac{17}{35} = \frac{7}{17}.$

You might wonder why the answer to (a) was not

$$\frac{\text{no. of girls}}{\text{no. of pupils}} = \frac{12+16}{30+28} = \frac{28}{58} = \frac{14}{29}.$$

Why does this argument give the wrong answer?

The reason this method does not produce the correct answer here is that the pupils are not all equally likely to be chosen. Each pupil in 7C has a probability of

$$\frac{1}{2}\times\frac{1}{30} = \frac{1}{60}$$

of being selected, but for those in 7K it is

$$\frac{1}{2}\times\frac{1}{28} = \frac{1}{56}.$$

Exercise 1E

1. Two cards are drawn successively from an ordinary pack of 52 playing cards and kept out of the pack. Find the probability that:

   (a) both cards are hearts;

   (b) the first card is a heart and the second card is a spade;

   (c) the second card is a diamond, given that the first card is a club.

2. A bag contains four red counters and six black counters. A counter is picked at random from the bag and not replaced. A second counter is then picked. Find the probability that:

   (a) the second counter is red, given that the first counter is red;

   (b) both counters are red;

   (c) the counters are of different colours.

3. The two events A and B are such that
   $$P(A)=0.6,\ P(B)=0.2,\ P(A|B)=0.1.$$
   Calculate the probabilities that:
   (a) both of the events occur;
   (b) at least one of the events occurs;
   (c) exactly one of the events occurs;
   (d) B occurs, given that A has occurred.

4. In a group of 100 people, 40 own a cat, 25 own a dog and 15 own a cat and a dog. Find the probability that a person chosen at random:
   (a) owns a dog or a cat;
   (b) owns a dog or a cat, but not both;
   (c) owns a dog, given that he owns a cat;
   (d) does not own a cat, given that he owns a dog.

1.7 Independence

In the previous section the answer to, "What is the probability that it will rain next Monday?" depended on the fact that you were told or knew about the season.

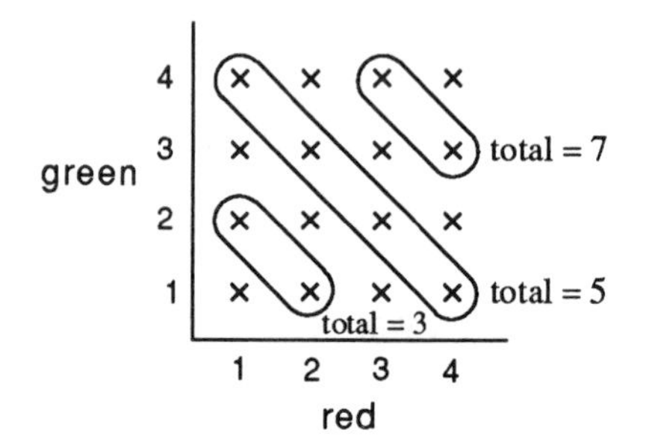

When two tetrahedral dice are rolled there are sixteen possible outcomes as shown in the diagram opposite.

What is $P(\text{total}) = 7$?

Now if I tell you that my cat has a broken leg, what is

$$P\ (\text{total } = 7 \mid \text{my cat has a broken leg})?$$

The answer is $\frac{2}{16}$ to both of these questions. The replies are the same because the two things discussed, the chance of a total of 7 and my cat having a broken leg, are independent. Other examples may not be as immediately obvious.

Example

What is the value of

(a) $P(\text{total} = 5)$ (b) $P(\text{total} = 5 \mid \text{red} = 2)$

(c) $P(\text{total} = 3)$ (d) $P(\text{total} = 3 \mid \text{red} = 2)$?

Solution

(a) From the sample space diagram above

$$P(\text{total} = 5) = \frac{4}{16} = \frac{1}{4}.$$

(b) Again $P(\text{total} = 5 \mid \text{red} = 2) = \dfrac{1}{4}$

since there is only one event (red = 2, green = 3) out of four possible events for red = 2.

(c) $P(\text{total} = 3) = \dfrac{2}{16} = \dfrac{1}{8}.$

(d) $P(\text{total} = 3 \mid \text{red} = 2) = \dfrac{1}{4}.$

The answers to (a) and (b) are both $\frac{1}{4}$, so the answer to ,"How likely is a total of 5?", is independent of (not affected by) the fact that you were told in (b) that the red score was 2.

(c) and (d) have different answers, however, $\frac{1}{8}$ and $\frac{1}{4}$ respectively, so your assessment of how likely a total of 3 is depends on the fact given in (d).

If two events, A and B, are such that

$$P(A \mid B) = P(A)$$

then they are said to be **independent**. Otherwise they are **dependent.**

In Section 1.5 there were examples of both cases. The tree diagram showing how pupils travelled to school included

$$P(\text{walk} \mid \text{boy}) = \frac{1}{4}$$

and $$P(\text{walk} \mid \text{girl}) = \frac{1}{10},$$

so how likely you think a pupil is to walk would depend on their sex.

On the other hand, in another example in Section 1.5, the chance of Jo scoring was not related to how likely Sam was to score so these events were independent. (In a tree diagram to show two events the branches are duplicated after each initial one if the second event is independent of the first.)

Example

In one year at school, 25 out of 154 failed the end of term maths exam. One class was particularly badly behaved and 7 out of 31 of them failed. Does bad behaviour in class affect how likely a pupil is to fail the test?

Solution

$$P(\text{fail}) = \frac{25}{154} = 0.162 \text{ (to 3 d.p.)}$$

$$P(\text{fail} \mid \text{badly behaved class}) = \frac{7}{31} = 0.226 \text{ (to 3 d.p.).}$$

Since these are certainly different the events are dependent, so the answer is 'Yes'.

Example

A family has three children. What is the probability that all three are the same sex? If you know at least two of them are girls what is the probability that they are all the same sex? Has this piece of information been of any help?

Solution

The possible combinations are shown below.

At least two girls	G	G	G	all same sex
At least two girls	G	G	B	
At least two girls	G	B	G	
	G	B	B	
At least two girls	B	G	G	
	B	G	B	
	B	B	G	
	B	B	B	all same sex

$$P(\text{all same sex}) = \frac{2}{8} = \frac{1}{4}$$

$$P(\text{all same sex} \mid \text{at least 2 girls}) = \frac{1}{4}.$$

So the events are independent, and the answer is 'No'.

Starting from the definition of independence,

$$P(A) = P(A \mid B)$$

$$= \frac{P(A \cap B)}{P(B)}$$

$$\Rightarrow \quad P(A)\,P(B) = P(A \cap B).$$

Testing to see whether or not $P(A) \times P(B)$ is, in fact, equal to $P(A \cap B)$ can also be used as a test for independence. So in our last example,

$$P(\text{at least two girls}) = \frac{4}{8} = \frac{1}{2}$$

$$P(\text{all three the same sex}) = \frac{2}{8} = \frac{1}{4}$$

$$P(\text{at least two girls} \cap \text{all three the same sex}) = \frac{1}{8}.$$

Since $\frac{1}{2} \times \frac{1}{4} = \frac{1}{8}$ you can see that these events are independent.

If A and B are independent then the occurrence of B does not affect the likelihood of A happening and similarly it seems very likely that the non-occurrence of B should have no effect.

If A and B are independent, then

$$P(A \cap B) = P(A)\,P(B)$$

$$P(A) - P(A \cap B) = P(A) - P(A)\,P(B)$$

$$\Rightarrow \quad P(A \cap B') = P(A) - P(A)\,P(B)$$

$$= P(A)\left[1 - P(B)\right]$$

$$\Rightarrow \quad P(A \cap B') = P(A)\,P(B').$$

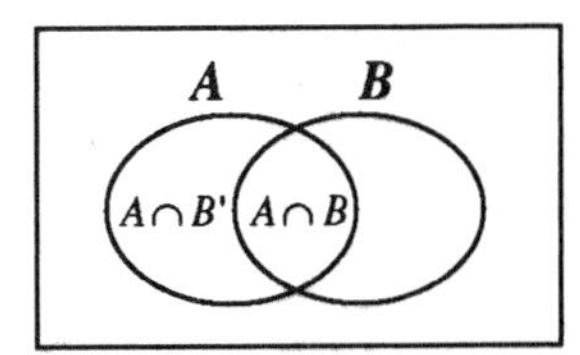

So A and B' are also independent.

Exercise 1F

1. A card is selected at random from an ordinary pack of 52. If

   A = the card is an ace
   D = the card is a diamond
   P = the card is a picture (Jack, Queen or King)
   R = the card is from a red suit
   X = the card is not the three of diamonds or the two of clubs,

   what are the values of the following:

   (a) $P(A)$ (b) $P(A|D)$ (c) $P(D)$
   (d) $P(D|P)$ (e) $P(D|R)$ (f) $P(P)$
   (g) $P(P|A)$ (h) $P(P|A')$ (i) $P(A|X)$
   (j) $P(D|X')$ (k) $P(X|D)$ (l) $P(R|X)$?

2. Which of the following pairs of events from Question 1 are independent:

   (a) A, D (b) D, P (c) P, A
   (d) R, X (e) D, R (f) D, R'?

3. Work out the six probabilities on the branches labelled a to f and also the value of g.

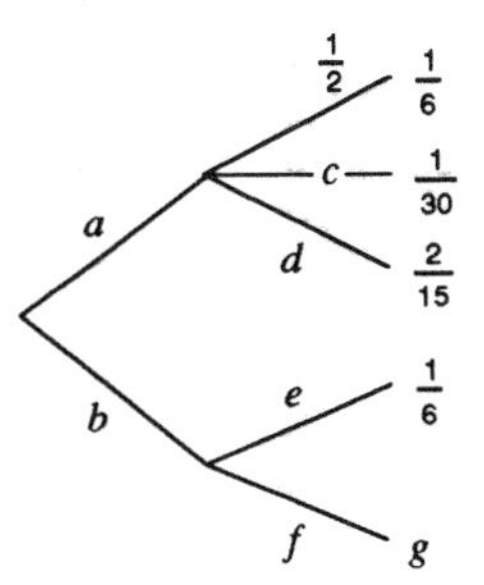

4. Two boxes, A and B, each contain a mixture of black discs and white discs. A contains 8 black and 7 white while B has 5 black and 7 white. A box is selected at random and a disc taken from it.

   Draw a tree diagram and calculate the probability that:

   (a) the disc is white;

   (b) the disc came from B if it is white.

5. A box contains 60 balls each of which is either red, blue or white. If the numbers of these are in the ratio 6:3:1 how many are there of each?

   By drawing a tree diagram, find the probability that when two balls are drawn at random together:

   (a) they are the same colour;

   (b) no red ball is drawn;

   (c) they are both white if you are told they are the same colour.

6. In a quiz competition the first question is worth one point and answered correctly with probability $\frac{5}{6}$. After any question is answered correctly the contestant receives one worth a point more than the previous one. After a wrong answer the contestant receives a one point question and two consecutive wrong answers eliminates the competitor.

   If the probabilities of correctly answering 2, 3 and 4 point questions are $\frac{4}{5}$, $\frac{3}{4}$ and $\frac{2}{3}$ respectively, calculate the probability that after four rounds the contestant has:

   (a) been eliminated;

   (b) scored at least six points.

7. In May, three mornings out of every five on average are fine.

   When the weather is fine Sarita walks to work with probability $\frac{1}{2}$, goes on the bus with probability $\frac{1}{3}$ and drives when she does not walk or use the bus. She never walks when the weather is wet and drives four times as often as she uses the bus. The probability of her arriving late when she walks in is $\frac{1}{2}$ and when she uses the bus it is $\frac{1}{4}$. She is always on time when she drives.

   On a particular May morning what is the probability that Sorita:

   (a) arrives on time;

   (b) travels by bus and is late?

   If she arrives late what is the probability that it is a fine morning?

1.8 Miscellaneous Exercises

1. One die has the numbers 1, 2, 3, 4, 5, 6 on its faces and another has 1, 1, 2, 2, 3, 3 on its faces. When the two are rolled together what is the probability that one of the scores will divide exactly into the other?

2. There are prizes for the first two runners home in a race with six competitors. What is the probability that:

   (a) both Dave and Raj will win prizes;

   (b) neither Dave nor Raj will win a prize?

3. A two digit number is written down at random. What is the probability that it:

   (a) is divisible by 5; (b) is divisible by 3;

   (c) is greater than 50; (d) is a square number?

4. When four coins are tossed together, what is the probability of at least three heads?

5. A counter starts at the point (0, 0). A coin is tossed and when a tail results it moves one unit to the right. When a head is seen it moves one unit upwards.

   What is the probability that after three goes it is still on the x-axis?

6. **Buffon's Needle** Take some ordinary pins and draw a set of straight lines across a sheet of paper so that they are the same distance apart as the length of a pin. Drop ten pins onto the lined paper several times and record your results in the same way as in Section 1.2, noting how many lie across a line. Draw a graph and estimate the probability of a pin crossing a line when dropped.

7. **Off-centre spinner** Make a hexagonal spinner and put a cocktail stick or something similar through a point to divide AB in the ratio 2 : 3.

   What are the probabilities of the various scores?

8. Four unbiased coins are tossed together. For the events A to D below, say whether statements (a) to (d) are true or false and give a reason for each answer.

   A = no heads B = at least one head

   C = no tails D = at least two tails

(a) A and B are mutually exclusive;

(b) A and B are exhaustive;

(c) B and D are exhaustive;

(d) A' and C' are mutually exclusive.

9. In a class of 30 pupils, 17 have a dog, 19 have a cat and 5 have neither. If a member of the form is selected at random what is the probability that this pupil has both a cat and a dog?

10. The probability that Suleiman will be chosen to play in goal for the next match is $\frac{1}{4}$ and the probability that Paul will be selected for that position is $\frac{2}{5}$. Find the probability that:

    (a) Suleiman or Paul will be selected to play in goal;

    (b) neither of them will be asked to play in goal.

11. A number is to be formed by arranging the digits 1, 4, 7 and 8 in some order.

    If A = the number is odd

    and B = the number is greater than 4000,

    find the value of:

    (a) $P(A)$ (b) $P(B)$ (c) $P(B|A)$

    (d) $P(A \cap B)$ (e) $P(A'|B)$.

12. John does $\frac{3}{5}$ of the jobs that come into the workshop and Dave does the rest. If 35% of John's work and 55% of Dave's work is perfect, find the probability that a job done in the workshop will be done:

    (a) perfectly;

    (b) by Dave if it was not done perfectly.

13. A warehouse receives 60% of its supplies from factory A, 30% from B and the rest from C.

    A sends large, medium and small items in the ratio 1 : 3 : 2.

    B's supplies are $\frac{1}{3}$ large size and no small size.

    C provides three times as many medium as small items and no large ones.

    If an item is selected at random from the warehouse, what is the probability that it is:

    (a) medium;

    (b) from B and large;

    (c) from C if it is found to be medium?

14. A box contains 8 discs of which 5 are red and 3 are blue. One is selected at random and its colour noted. It is returned to the box together with an extra one of the other colour. This process is repeated twice more.

    What is the probability that:

    (a) the third disc selected is red;

    (b) more reds are selected than blues;

    (c) the third disc is red if there are more blues shown than reds?

15. At a fete, one of the games consists of throwing a 2p coin onto a large board of coloured squares, each 2 inches by 2 inches. If a coin lies completely within a single square it is returned to a player with another 2p, otherwise it goes to the organiser. A 2p coin has a diameter of 1 inch. By considering where the centre of the coin must land for a win, work out the player's probability of success.

    How much money should the organiser expect to take in one hundred goes?

    To make more profit, you could draw up a board to use with 10p coins. What size should the square be if the player is to have a probability of $\frac{1}{3}$ of winning? (Answer to the nearest mm.)

16. A circular spinner has three sections numbered 1, 2 and 3. If these numbers came up twenty-five, thirty and forty-five times in an experiment, what do you consider the likely values for the angles of the sectors?

17. Twenty discs numbered 1 to 20 are put at random into four boxes with five in each.

    What is the probability that numbers 15 and 19 will be in the same box?

    Would the answer be different if the discs had been split into five groups of four?

18. A forgetful teacher leaves his mark book in a room where he has had a lesson once in every three occasions on average. If he teaches three lessons in different rooms in the morning, what is the probability that:

    (a) he will arrive at the lunch break having lost his mark book;

    (b) he left it in the second room if he finished the morning without it?

19. Three bags, A, B and C, each contain three 5p coins and two 2p coins. A coin is selected at random from A and placed in B. A coin is drawn from B and put in C. Finally, a coin is drawn from bag C. Find the probability that:

    (a) the coin selected from C is a 2p;

    (b) the coin selected from A was a 5p if the one from C was a 2p.

    Explain why the answer to (a) might have been expected. Repeat (a) for x 2p coins and y 5p coins, showing all your working.

20. Four girls each try to catch a ball and the probability that each will succeed is independently $\frac{2}{3}$.

    What is the probability that it will:

    (a) not be caught;

    (b) be caught?

21. Three students, Dave, Jane and Mary, share a house. Each of the girls is twice as likely as Dave to receive a telephone call in the evening. The probabilities that each will be out on any evening are independently $\frac{1}{2}$, $\frac{2}{5}$ and $\frac{3}{5}$ respectively. If the telephone rings one evening find the probability that the call is:

    (a) for Jean who is in;

    (b) for someone who is out;

    (c) for Dave given that it is for someone who is out.

22. Two gamblers play a game with two coins. The first tosses them and pays the second £1 for each head showing. Then the second has a turn and pays £1 for each tail showing. After each has had one go what is the probability that the first player has made a profit?

23. A school has three minibuses and the probability that each is free after school is independently $\frac{2}{5}$.

    Find the probability that after school on a particular day:

    (a) at least one minibus is free;

    (b) all the minibuses are free if at least one is free.

24. A set of dominoes consists of twenty eight pieces, each of which shows two sets of spots from zero to six, and no two dominoes are the same. A single domino is selected at random. Show the 28 possibilities on a diagram.

    What is the probability that:

    (a) the smaller number is 2;

    (b) it is a double;

    (c) it contains neither a 4 nor a 5?

25. Three coins are tossed. Event X is that at least one head and at least one tail result. Event Y is that at most one head shows. Are events X and Y independent?

26. Vehicles approaching a crossroad must go in one of three directions - left, right or straight on. Observations by traffic engineers showed that of vehicles approaching from the north, 45% turn left, 20% turn right and 35% go straight on. Assuming that the driver of each vehicle chooses direction independently, what is the probability that of the next three vehicles approaching from the north

    (a) (i) all go straight on;

    (ii) all go in the same direction;

    (iii) two turn left and one turns right;

    (iv) all go in different directions;

    (v) exactly two turn left?

    (b) Given that three consecutive vehicles all go in the same direction, what is the probability that they all turn left? (AEB)

27. (a) A student has a fair coin and two six-sided dice, one of which is white and the other one blue. The student tosses the coin and then rolls both dice. Let X be a random variable such that if the coin falls heads, X is the sum of the scores on the two dice, otherwise X is the score on the white die only.

    Find the probability function X in the form of a table of X and their associated probabilities.

    Find $P(3 \le X \le 7)$.

    State the assumption you made to enable you to evaluate the probability function.

    (b) The results of a traffic survey of the colour and type of car are given in the following table.

	Saloon	Estate
White	68	62
Green	26	32
Black	6	6

    One car is selected at random from this group.

    Find the probability that the selected car is

    (i) a green estate car,

    (ii) a saloon car,

    (iii) a white car given that it is not a saloon car.

    Let W and G denote the events that the selected car is white and green respectively and let S be the event that the car is a saloon.

    Show that the event $W \cup G$ is independent of the event S. Show, however, that colour and type of car are not independent. (AEB)

28. The staff employed by a college are classified as academic, administrative or support. The following table shows the numbers employed in these categories and their sex.

	Male	**Female**
Academic	42	28
Administrative	7	13
Support	26	9

A member of staff is selected at random.

A is the event that the person selected is female.

B is the event that the person selected is academic staff.

C is the event that the person selected is administrative staff.

($\overline{A}$ is the event not A, $\overline{B}$ is the event not B, $\overline{C}$ is the event not C)

(a) Write down the values of

(i) $P(A)$,

(ii) $P(A \cap B)$

(iii) $P(A \cup \overline{C})$

(iv) $P(\overline{A} | C)$

(b) Write down one of the events which is

(i) not independent of A,

(ii) independent of A,

(iii) mutually exclusive of A.

In each case, justify your answer.

(c) Given that 90% of academic staff own cars, as do 80% of administrative staff and 30% of support staff,

(i) what is the probability that a staff member selected at random owns a car?

(ii) A staff member is selected at random and found to own a car. What is the probability that this person is a member of the support staff?

(AEB)

29. A vehicle insurance company classifies drivers as A, B or C according to whether or not they are a good risk, a medium risk or a poor risk with regard to having an accident. The company estimates that A constitutes 30% of drivers who are insured and B constitutes 50%. The probability that a class A driver will have one or more accidents in any 12 month period is 0.01, the corresponding values for B and C being 0.03 and 0.06 respectively.

(a) Find the probability that a motorist, chosen at random, is assessed as a class C risk and will have one or more accidents in a 12 month period.

(b) Find the probability that a motorist, chosen at random, will have one or more accidents in a 12 month period.

(c) The company sells a policy to a customer and within 12 months the customer has an accident. Find the probability that the customer is a class C risk.

(d) If a policy holder goes 10 years without an accident and accidents in each year are independent of those in other years, show that the probabilities that the policy holder belongs to each of the classes can be expressed, to 2 decimal places, in the ratio 2.71 : 3.69 : 1.08. (AEB)

30. A hospital buys strawberry jam in standard sized tins from suppliers A, B and C. (The table on the next page gives information about the contents.)

Find the probability of a tin selected at random being

(a) from supplier A,

(b) underweight.

What is the probability of

(c) a tin from B being both underweight and poor quality,

(d) an underweight tin from A containing poor quality jam,

(e) a tin from C being both underweight and poor quality,

(f) a tin from C which contains poor quality jam being underweight,

(g) a tin selected at random being both underweight and poor quality.

(h) a tin being from A given that it is both underweight and of poor quality?

(AEB)

Supplier	% of hospital requirements supplied	% of tins with underweight contents	% of tins containing poor quality jam	Other information
A	55	3	7	1% are both underweight and poor quality
B	35	5	12	probability of poor quality is independent of probability of being underweight
C	10	6	20	40% of underweight tins contain poor quality jam

2 DATA COLLECTION

Objectives

After studying this chapter you should

- understand what is meant by qualitative and quantitative data, discrete and continuous variables;
- understand what is meant by primary and secondary data;
- use random number tables to find samples;
- be able to find random, systematic, stratified, quota and cluster samples.

2.0 Introduction

The current 'life expectancy' in the UK is about 71 years for men and 77 years for women. Apart from the obvious interest to individuals, figures such as these are of great concern to others: insurance companies, health organisations, social services, government departments such as the Treasury, leisure companies, etc. This kind of information is therefore collected by the government by means of the census and other surveys. A census is usually carried out every 10 years in this country and is compulsory by law to complete. Before modern technology was available it took several years to analyse the results, by which time much of the information was out of date anyway. In this chapter you will meet some of the techniques which might be used in such an analysis.

Consider the following two questions:

If you were told that your blood pressure was 140/90 would this be normal?

What is the normal weight of a seventeen-year-old in kilograms?

These are typical of the types of questions to be answered. You will need about 30 people for the first activity so you may have to involve other groups. You may not be able to carry out all the tests suggested in the following Activity but do try to obtain some of the equipment to do the more interesting and unusual ones - most of it is probably available in your school or institution if you ask. Do check that you know how to use the equipment properly.

Activity 1

In all tests your subjects should be allowed to test themselves. Keep all results confidential. Record, however, whether each participant was male or female. This Activity involves gathering data and you will be expected to analyse the data later in this chapter.

Measure the following:

(a) The **heights** in cm and **weights** in kg of everyone. Two metre rules taped to the wall and a book on the head works best for height. Weight is most easily measured by bathroom scales.

(b) **Eye** and **hair colour**. Make sure hair colour is natural! Decide on the categories before you start and stick to these.

(c) The number of occasions in the last month that individuals have undertaken hard **physical exercise** lasting 20 minutes or more, e.g. hockey, swimming, cycling to school.

(d) **Blood pressure**. Cheap digital blood pressure meters are available on the market and many Biology/P.E. Departments have these. Blood pressure is measured in two ways:

(i) **Systolic** - taken when the heart is beating and exerting maximum pressure.

(ii) **Diastolic** - taken when the heart is at rest and pressure is at minimum.

These are usually written together, e.g. 120/60. Take both these readings.

(e) **Pulse**. Digital blood pressure machines usually give this as well. If not, rather than use the traditional pulse point on the wrists, it is often easier to measure it with two fingers on the side of the throat. Count the beats in half a minute and double the result.

(f) **Breath power**. Blowmeters are commonly held by medical centres as they are useful in assessing asthmatics. Your Biology or PE Department may have one. By blowing into them the lung capacity can be measured.

(g) **Reaction times**. Reaction rulers are commercially available which can be used to measure your reactions. Alternatively, take a ruler marked in centimetres and hold it above the subject's slightly opened thumb and forefinger so that these are level with the zero on the ruler. When the ruler is dropped, the subject catches it. Measure the distance (in centimetres) the ruler drops before it is caught.

2.1 What sort of data?

The data on the next page give various information on share prices on the London Stock Exchange. Data which you have collected yourself are called **primary** data, but data such as the Stock Market publish, where you are relying on someone else's measurements, are a **secondary** source.

Activity 2 Primary and secondary sources

Working in small groups discuss the following questions:

In each of these cases what possible sources of secondary data might be available? How might a survey be carried out? What are the advantages and disadvantages of using primary or secondary sources?

(a) The Health Education Council wants to know if a new campaign to stop young people starting smoking has been effective.

(b) A school canteen wants to see if there is a demand for healthier foods.

(c) A scientist wants to measure if a low fat diet improves athletic performance.

An even more important distinction between types of data is to what extent numbers are involved.

Qualitative data is where the actual measurements have no meaningful value, e.g. starting letter of Stock name, colour of a company logo. Be careful, as sometimes when recording data codes are used, e.g. 0 for male, 1 for female.

Quantitative data is where the data has a valid numerical value, e.g. share price. This category is further subdivided into

(a) **discrete** - where the data can only be one of a fixed number of numerical values, usually, but not necessarily, whole numbers, e.g. change on week.

(b) **continuous** - where the data can fall anywhere over a range and the scale is only restricted by the accuracy of measuring, e.g. yield (these are rounded to 1 d.p.).

Sometimes the division between discrete and continuous is a little indistinct. For example, share prices are strictly speaking discrete since they can only be to the nearest $\frac{1}{2}$p but because of the wide range of values it would be far more convenient to regard them as continuous.

London: The FT-SE 100

Stock	Price	Change	Yield
Abbey National	274	-3	4.6
Allied - Lyons	554	+5	4.5
Anglian Water	286	-8	6.8
Argyll Group	305	0	3.9
Arjo Wiggins Teape	252	+1	4.4
Asda Group	105	-8	6.1
Ass Brit Foods	534	+8	3.0
BAA	436	+7	3.5
Bank of Scotland	104	0	6.5
Barclays Bank	432	0	6.5
Bass	967	-7	4.5
BAT Inds	732	+11	5.7
BET	181	+14	9.6
BICC	440	+2	5.8
Blue Circle Inds	243	+2	6.2
BOC Group	562	+14	4.8
Boots	397	+14	4.0
British Aerospace	587	-11	5.7
British Airways	172	+3.5	6.9
British Gas	250	-1	6.9
BP	334	-2	6.6
British Steel	135	+0.5	8.1
British Telecom	381	+3	4.8
BTR	395	+3	5.3
Cable & Wireless	547	+42	2.9
Cadbury Schweppes	352	-13	4.4
Commercial Union	491	+17	6.2
Courtaulds	402	+12	4.0
Enterprise Oil	513	-13	3.9
Eurotunnel Units	470	+7	–
Fisons	494	+7	2.0
Forte	271	+3	4.9
General Accident	528	+8	6.8
GEC	192.5	-1	6.4
Glaxo Holdings	1280	+42	2.3
Grand Metropolitan	771	+12	3.6
Gt Universal Stores	1228	+32	3.7
GRE	199	+4	8.0
Guinness	985	+25	2.5
Hammerson 'A'	608	+2	4.5
Hanson	216.5	-5.5	6.5
Harrisons & Cros	148	+4	8.1
Hawker Siddeley	581	+9	5.7
Hillsdown Holdings	228	-4	4.7
ICI	291	-11	5.7
Kingfisher	499	+4	3.3
Ladbroke	268	+9	5.3
Land Securities	503	0	5.2
Lasmo	327	-12	3.5
Legal & General	433	+18	5.5
Lloyds Bank	338	+5	6.0
Lonrho	243	0	8.8
Lucas Inds	154	+3	6.1
Marks & Spencer	253	+4	3.5
Maxwell Comm	207.5	0	10.0
MEPC	474	-2	5.3
Midland Bank	211	-3	5.3
Nat Power	141	0	5.1
NatWest	313	+4	7.5
NW Water	288	+3	6.9
Pearson	730	+5	4.2
P&O dfd	572	+2	7.1
Pilkington	178	+5	8.2
Powergen	147.5	0	5.0
Prudential Corp	237	+3	5.8
Racal Electronics	221	-20	2.3
Rank Org	685	-4	6.0
RHM	270	-5	6.3
Reckitt & Coleman	1580	+2	2.9
Redland	561	-5	5.9
Reed International	432	+29	4.7
Reuters	824	+5	2.4
RMC Group	657	-16	3.9
Rolls - Royce	155	-7	6.2
Rothmans	914	+21	2.2
Royal Bank of Scotland	180	-1	6.2
Royal Insurance	436	+14	8.0
RTZ	550	-5	4.7
Sainsbury	374	+5	2.6
Scottish & Newcastle	393	+4	4.4
Sears	78	-4	9.2
Severn Trent	254	-4	6.1
Shell Transport	514	+1.5	5.2
Smith Kline Beecham	781	-6	2.4
Smith & Nephew	134.5	-0.5	4.3
Sun Alliance	370	+11	5.0
Tarmac	224	-9	6.7
Tate & Lyle	390	+35	3.4
Tesco	278	-1	2.5
Thames Water	292	-6	6.6
Thorne EMI	739	+3	5.7
Trafalgar House	256	+6	9.6
TSB	147	+1.5	5.8
Ultramar	287	-6	4.9
Unilever	745	-10	3.3
United Biscuits	361	-4	5.3
Wellcome	643	+14	1.3
Whitbread 'A'	500	-7	4.3
Williams Hldgs	308	+11	5.2
Willis Corroon	302	+12	5.8

Activity 3

Make a list of all the information you measured in Activity 1 and classify it under the three types of data.

2.2 Sources of data

The UK government produces vast quantities of statistical information in its many departments. These are mainly coordinated by two main bodies:

Central Statistics Office - largely responsible for producing and checking all the information produced by individual departments;

Office of Population Censuses and Surveys - carry out any data collection based on the general public. In particular the Census is carried out every 10 years and the Household Expenditure Survey is carried out on an ongoing basis.

The address of these departments is given along with other useful addresses in the Appendix.

One essential publication to have is:

Government Statistics -A brief guide to sources. This is obtainable from your local HMSO supplier. It contains a list of all the important publications produced by the government and details of how to obtain them. The most useful of these are shown below and may be available from your library or from HMSO suppliers.

General digests

Monthly Digest of Statistics
Collection of main series from all government departments.
Monthly.

Annual Abstract of Statistics
Contains many more series than the *Monthly Digest* and provides a longer run of years.
Annual.

Key Data
Contains over 130 tables, maps and coloured charts and covers a wide range of social and economic data. Each table and chart is accompanied by a reference to sources.
Annual.

Social Trends
Brings together key social and demographic series in colour charts and tables.
Annual.

Regional Trends
A selection of the main statistics that are available on a regional basis.
Annual.

The Annual Abstract and Social Trends are a mine of information in many fields and are kept by all good reference libraries.

In addition to the periodical data collections used in the above, various one-off reports are commissioned by the government. Examples are:

Skateboarding Accidents in the UK - a report on accidents involving people using skateboards giving information on the nature of the accidents and injuries sustained.

Smoking/drinking amongst schoolchildren. Several studies have been carried out in these areas.

Heights and weights of people. Broken down into different age groups, for example you can find the distribution of heights and weights for 16-19 year olds in the country as a whole.

As well as the UK government sources there are a number of other international bodies that produce statistical information. Catalogues of available publications can be obtained from your local HMSO free of charge. Some useful sources of information are:

European Community - produces much Annual Abstract/Social Trends-type data for countries in Europe. In particular, *Europe in Figures* is an inexpensive book produced annually. In addition there are a great number of reports on different issues such as employment, women's rights and the environment.

UNESCO (United Nations Educational Scientific & Cultural Organisation) - produces many publications in its field, not all statistical.

WHO (World Health Organisation) - much of it fairly technical but some interesting reports on smoking/alcoholism.

Other UK institutions providing data include:

Association of British Insurers - produce statistical information on all aspects of insurance.

Building Societies Association - in particular produces regular 'bulletins' with information on regional house prices.

High Street banks - produce regular reviews in addition to various economic and business data for their customers.

Banking Information Service - provides information about UK banks.

Market Research Society - in particular the *MRS Yearbook* contains useful tables on 'Market penetration of durable goods' on a regional basis. Also it has useful information on how to carry out surveys.

Meteorological Office - produces summary statistics on weather.

Various directories of business information exist giving details of companies' activities and important financial information. Company reports/share prospectuses give information in the notes to the accounts.

It should also be noted that the quality newspapers make frequent use of statistics in articles, as well as regularly publishing statistics, particularly financial. Other periodicals in fields such as economics, sociology, etc. have regular features that use statistics.

Activity 4

Take **one** of the following topics as an investigation. Collect as much information as you can using the above sources or any others you can find. Try to find at least three different sources. Write a short report using the information as reference. Outline what primary information you might collect locally for further investigation.

(a) Does the legal age of drinking/smoking need to be lowered in view of the fact that many under-age youngsters already partake?

(b) Has current government economic policy enabled small businesses to survive more easily?

(c) Has the AIDS publicity in the early 1990s promoted a more responsible attitude towards sex in young people?

(d) Has the greater awareness of environmental issues in recent years led to any noticeable improvements in the way we look after the environment?

2.3 Sampling: factors and bias

You will have seen that secondary data can be extremely useful in investigations and will probably be collected on a much grander scale than can be done at your level. However, frequently you will be working in a new area or wish to collect your own data locally.

Every 10 years (since 1801) the OPCS (Office of Population Census and Surveys) carries out a census for the government. The word **census** means to include everybody.

The article on the following page shows the scale of such a piece of work.

ON Sunday all householders in England, Scotland and Wales will have to fill out a form giving details of everyone who lives at their address as part of the 19th full British census.

A census is a national survey to count the population and collect information which government departments will use to plan policies. The census will attempt to give a picture of Britain at midnight on April 21. People who use the figures will be able to compare the results with statistics collected in previous censuses to find out how Britain's population and society are changing.

A 12-page form is being delivered to, and will be collected from the country's 23 million households by people known as "numerators". There are about 115,000 of these specially recruited temporary staff. Each is responsible for about 200 households.

A further 1, 800 temporary staff will key the census information into a massive government computer in Titchfield, Hampshire. The whole process of collecting and processing the data costs the Government about £135 million.

In this year's census new questions will be asked about people's ethnic origin and any long-term illness they might have. For the first time, an attempt will be made to count the number of homeless people in Britain.

The census is held every ten years on a Sunday, the day most people are at home. It is organised by the Office of Population Censuses and Surveys (OPCS) in England and Wales, and by the Registrar General in Scotland. Separate censuses will also be held on April 21 in Northern Ireland and the Irish Republic.

Most countries count their populations. The United States, for instance, has held a census every 10 years since 1790. Early this year a census in India showed that it has a total population of 844 million people. Australia's latest census, by contrast, showed it has just 17 million people spread across a land area twice as large as India's.

In 1975 the government wanted census information before the 1981 full census, so the OPCS carried out a ten per cent census using 1 in 10 of the population. This is known as a **survey**.Data are obtained by asking people to fill in forms which are then given to collectors trained to sort out any queries.

In a research project looking at the disappearance of vegetation on mountain moorland, a scientist chose three specific sites to investigate. Fifty samples were selected at each site using a device called a quadrat (a 10 cm wire square) thrown at random into the undergrowth. The number of species of each type and the sizes were noted by students who were able to identify the plants.

Both these examples illustrate the same principle. When deciding how to carry out a data collection there are several decisions to be made:

(a) What size of sample can you reasonably expect to take, given limited time, money and resources?
(b) How are the items to be used in the sample to be chosen to avoid introducing bias?
(c) How is the data to be collected to avoid any bias?

The answer to question (a) clearly depends on the individual circumstances. It should be obvious, however, that the larger the sample the more sensitive the result.

In questions (b) and (c) the key element is to eliminate possible bias. In order to understand **bias** the idea of **factors** in an experiment is important. You are usually interested in one or more factors and their effect. However, there will always be other factors which might affect the result. For example, a horticulturist wishes to test the effect of a new fertilizer on different varieties of wheat. Some possible factors affecting the experiment could be listed as:

Relevant Factors	Bias Factors
Whether fertilizer used	Type of soil
Strength of fertilizer	Weather conditions
Variety of wheat used	Quality of seeds
	Care of plants
	Measurement of crop
	Position in field

The strength of fertilizer is really a sub factor of whether a fertilizer is used or not. You could list the strength as litres per square metre including zero. These are called the **levels of a factor**.

Activity 5

Make a list of relevant and bias factors for these experiments :

(a) Testing a new fuel additive to improve mileage in different cars.

(b) Testing whether a new language laboratory improves student performance in modern and classical languages.

(c) Examining the effect of alcohol on men's and women's reaction time.

(d) Asking people's opinions of current unemployment.

Where there are levels of a factor, indicate possible values the levels could take.

If a firework manufacturer wanted to test whether his product worked he could not possibly test every item as he would have nothing left to sell. He would try to take a 'representative' sample of all the fireworks he produced. By **representative** we mean that the sample has approximately the same properties as the total 'population'. This is illustrated in the following case study.

A landowner has decided to sell a mature piece of deciduous woodland of 200 trees. He has asked a surveyor to come and assess the quality of the woods, but in the time available she can only carefully examine 50 trees. The landowner has a map of the woods (shown on the following page) on which he has numbered all the trees and indicated the variety. The surveyor says that the following details will be needed for each of 50 trees:

(a) the girth ;

(b) the age;

(c) whether it suffers from a major disease;

(d) the approximate height.

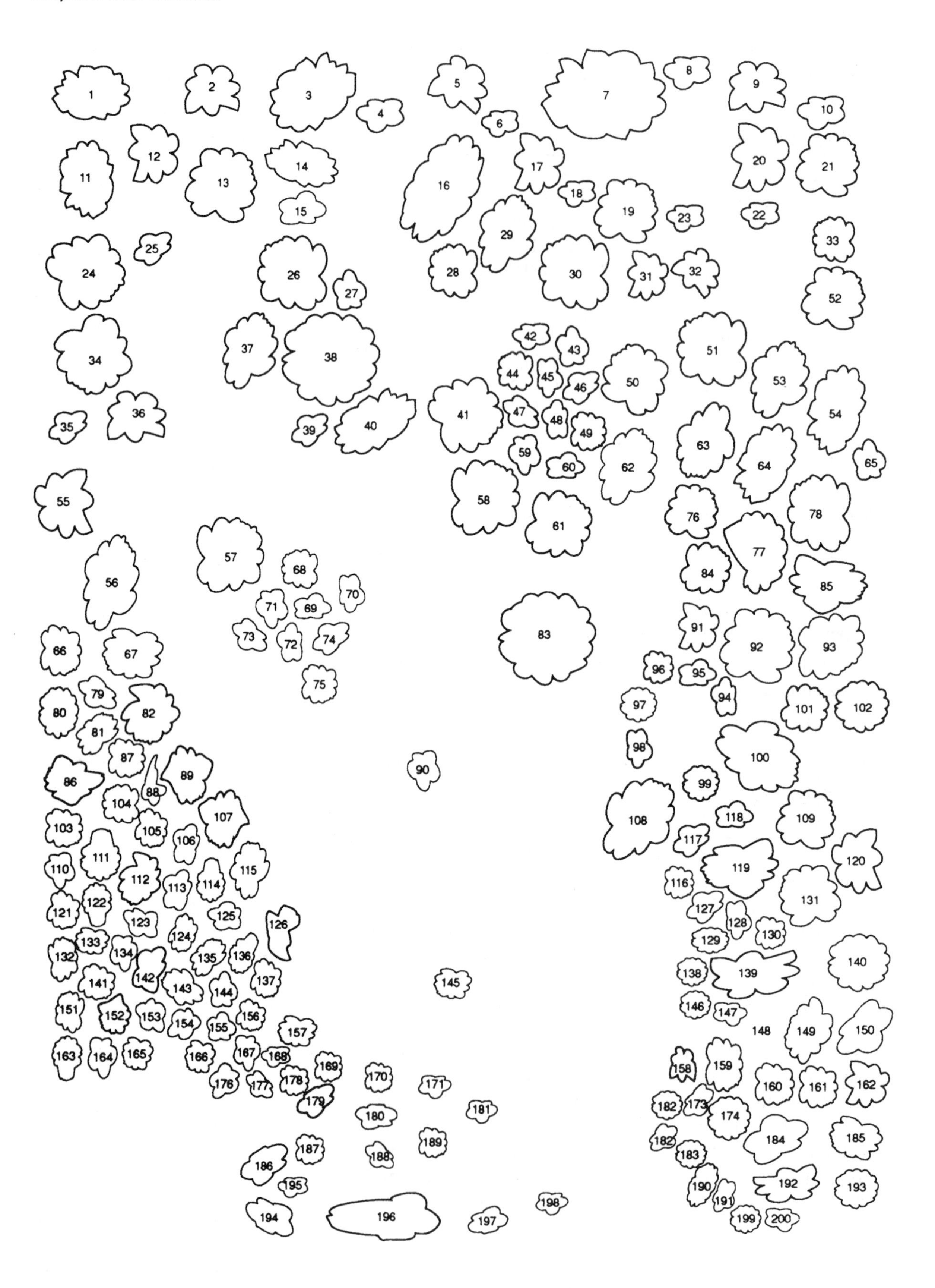

1
2
3
4
5
6
7
8
9
10
11
12
13
14
15
16
17
18
19
20
21
22
23
24
25
26
27
28
29
30
31
32
33
34
35
36
37
38
39
40
41
42
43
44
45
46
47
48
49
50
51
52
53
54
55
56
57
58
59
60
61
62
63
64
65
66
67
68
69
70
71
72
73
74
75
76
77
78
79
80
81
82
83
84
85
86
87
88
89
90
91
92
93
94
95
96
97
98
99
100
101
102
103
104
105
106
107
108
109
110
111
112
113
114
115
116
117
118
119
120
121
122
123
124
125
126
127
128
129
130
131
132
133
134
135
136
137
138
139
140
141
142
143
144
145
146
147
148
149
150
151
152
153
154
155
156
157
158
159
160
161
162
163
164
165
166
167
168
169
170
171
172
173
174
175
176
177
178
179
180
181
182
183
184
185
186
187
188
189
190
191
192
193
194
195
196
197
198
199
200

Tree	Type	Girth	Age	Disease	Height	Value
1	Oak	2.1	80	0	7	120
2	Elm	1.8	65	0	6	90
3	Oak	3.5	115	0	8	200
4	Birch	0.8	20	0	3	0
5	Elm	1.9	65	0	6	95
6	Birch	0.6	18	0	3	0
7	Oak	4.6	150	0	8	300
8	Birch	0.7	19	1	3	0
9	Elm	1.7	60	1	5	0
10	Birch	0.8	21	0	2	0
11	Elm	1.7	70	0	6	80
12	Elm	1.7	72	0	6	80
13	Oak	2.1	90	0	7	120
14	Yew	2.3	130	0	7	300
15	Birch	0.7	20	1	3	0
16	Oak	4.5	145	0	7	240
17	Elm	2.1	75	0	6	90
18	Birch	0.7	18	0	3	0
19	Oak	3.2	108	0	8	180
20	Elm	1.7	67	1	6	0
21	Elm	1.6	65	1	6	20
22	Birch	0.7	18	1	3	0
23	Birch	0.6	15	0	2	0
24	Oak	2.9	102	0	7	115
25	Birch	0.6	21	0	3	0
26	Oak	3.1	110	0	8	175
27	Birch	0.9	23	1	3	0
28	Elm	1.8	74	0	6	90
29	Oak	3.2	110	0	8	170
30	Oak	3.8	120	0	9	195
31	Elm	2.0	75	1	6	0
32	Elm	2.3	75	1	7	30
33	Elm	2.2	75	1	7	0
34	Elm	2.6	80	0	7	90
35	Birch	0.6	20	0	3	0
36	Elm	2.5	78	0	7	95
37	Elm	2.8	85	0	7	100
38	Oak	3.7	116	1	8	80
39	Birch	0.7	23	1	3	0
40	Elm	2.8	80	0	7	95
41	Elm	3.3	95	0	7	110
42	Birch	0.6	21	0	3	0
43	Birch	0.6	20	0	3	0
44	Birch	0.5	17	0	2	0
45	Birch	0.6	22	0	3	0
46	Birch	0.6	21	1	3	0
47	Birch	0.6	20	0	3	0
48	Birch	0.6	21	1	3	0
49	Birch	0.5	18	0	3	0
50	Elm	3.5	98	0	7	120
51	Oak	4.1	120	0	8	180
52	Oak	3.9	115	0	7	165
53	Oak	3.1	85	0	7	135
54	Oak	4.1	118	0	8	170
55	Elm	2.8	80	0	7	95
56	Oak	4.0	118	0	8	170
57	Yew	4.7	120	0	9	280
58	Elm	3.3	90	0	6	100
59	Birch	0.6	21	0	3	0
60	Birch	0.6	20	0	3	0
61	Elm	3.2	85	0	6	80
62	Elm	3.2	88	0	6	80
63	Oak	3.5	108	0	8	150
64	Oak	3.4	105	0	8	145
65	Elm	2.1	45	0	6	60
66	Beech	2.5	55	0	5	70
67	Oak	3.0	90	0	7	130
68	Birch	0.7	23	0	3	0
69	Birch	0.6	22	0	3	0
70	Birch	0.6	22	1	3	0
71	Birch	0.6	21	0	3	0
72	Birch	0.6	20	1	3	0
73	Birch	0.7	22	0	3	0
74	Birch	0.6	21	0	3	0
75	Birch	0.6	21	0	3	0
76	Elm	2.9	81	0	7	90
77	Oak	4.3	125	0	8	190
78	Oak	4.4	127	0	8	195
79	Beech	2.4	55	0	5	70
80	Beech	2.6	55	0	5	75
81	Beech	2.4	55	0	5	70
82	Oak	3.5	98	0	7	150
83	Yew	5.0	150	0	9	300
84	Elm	2.8	78	0	7	85
85	Oak	4.3	125	0	8	185
86	Beech	2.6	55	0	6	80
87	Beech	2.5	55	0	5	75
88	Beech	2.5	55	0	5	75
89	Oak	3.6	100	0	7	145
90	Beech	2.9	80	0	7	90
91	Elm	2.8	81	0	7	85
92	Oak	3.4	102	0	7	150
93	Oak	3.6	102	0	7	150
94	Birch	0.6	21	0	3	0
95	Birch	0.6	20	0	3	0
96	Birch	0.5	18	0	3	0
97	Birch	0.6	20	0	3	0
98	Birch	0.6	21	1	3	0
99	Birch	0.6	20	1	3	0
100	Elm	2.9	80	1	7	20

Tree	Type	Girth	Age	Disease	Height	Value
101	Elm	2.8	83	1	7	0
102	Elm	2.7	80	1	7	15
103	Beech	2.6	55	0	7	75
104	Beech	2.5	55	0	7	70
105	Beech	2.4	55	0	7	60
106	Beech	2.4	55	0	7	60
107	Oak	4.2	102	1	8	30
108	Elm	4.3	98	0	8	110
109	Elm	3.1	84	1	8	15
110	Beech	2.5	55	0	7	75
111	Beech	2.4	55	1	7	10
112	Beech	2.5	55	0	7	70
113	Beech	2.5	55	0	7	75
114	Beech	2.4	55	0	7	70
115	Oak	3.9	95	0	8	130
116	Birch	0.6	20	0	3	0
117	Birch	0.6	19	0	3	0
118	Birch	0.7	22	0	3	0
119	Yew	4.1	110	0	8	200
120	Elm	3.3	85	0	8	120
121	Beech	2.6	55	0	7	75
122	Beech	2.5	55	0	7	70
123	Beech	2.5	55	0	7	70
124	Beech	2.5	55	0	7	70
125	Beech	2.6	55	0	7	75
126	Oak	3.7	90	0	8	125
127	Birch	0.6	20	0	3	0
128	Birch	0.7	21	0	3	0
129	Birch	0.6	20	0	3	0
130	Birch	0.6	20	0	3	0
131	Elm	3.5	90	0	8	130
132	Beech	2.5	55	0	7	75
133	Beech	2.4	55	0	7	70
134	Beech	2.5	55	0	7	75
135	Beech	2.3	55	0	6	60
136	Beech	2.5	55	0	7	75
137	Beech	2.5	55	0	7	75
138	Birch	0.6	20	0	3	0
139	Beech	2.2	48	0	6	60
140	Elm	3.7	87	1	7	10
141	Beech	2.5	55	1	7	20
142	Beech	2.6	55	0	7	80
143	Beech	2.5	55	0	7	75
144	Beech	2.5	55	0	7	75
145	Beech	2.3	47	0	6	60
146	Birch	0.6	20	0	3	0
147	Birch	0.7	22	1	3	0
148	Oak	3.8	85	0	7	140
149	Oak	3.6	85	0	7	130
150	Oak	4.1	85	0	8	150
151	Beech	2.5	55	0	7	75
152	Beech	2.5	55	0	7	75
153	Beech	2.4	55	0	7	70
154	Beech	2.5	55	0	7	75
155	Beech	2.5	55	1	7	15
156	Beech	2.4	55	0	7	70
157	Elm	3.9	85	0	8	80
158	Birch	0.6	20	0	3	0
159	Oak	4.3	85	0	7	160
160	Oak	3.9	85	0	7	150
161	Oak	3.8	85	0	7	150
162	Oak	3.8	85	0	7	150
163	Beech	2.4	55	0	7	70
164	Beech	2.5	55	0	7	75
165	Beech	2.4	55	0	7	70
166	Beech	2.5	55	0	7	75
167	Beech	2.5	55	0	7	75
168	Birch	0.6	19	0	3	0
169	Birch	0.6	20	0	3	0
170	Birch	0.6	19	0	3	0
171	Birch	0.6	20	1	3	0
172	Birch	0.5	17	1	3	0
173	Birch	0.6	18	0	3	0
174	Elm	3.2	76	1	7	10
175	Beech	2.5	55	0	7	75
176	Beech	2.7	55	0	7	80
177	Birch	0.7	21	0	3	0
178	Birch	0.6	19	0	3	0
179	Beech	1.4	22	0	4	15
180	Birch	0.6	19	0	3	0
181	Birch	0.6	18	0	3	0
182	Birch	0.6	19	0	3	0
183	Birch	0.6	19	0	3	0
184	Elm	3.5	83	0	7	85
185	Elm	2.9	72	0	7	75
186	Beech	2.5	55	0	7	75
187	Birch	0.6	20	0	3	0
188	Birch	0.5	15	0	2	0
189	Beech	1.7	31	0	5	30
190	Beech	1.6	28	0	4	20
191	Birch	0.6	17	0	3	0
192	Elm	2.7	54	0	5	30
193	Elm	2.9	51	0	5	30
194	Elm	2.9	48	0	5	30
195	Birch	0.6	15	1	3	0
196	Yew	4.2	124	0	8	200
197	Beech	1.9	38	0	6	35
198	Birch	0.6	19	0	3	0
199	Birch	0.6	18	0	3	0
200	Birch	0.6	21	0	3	0

From this information it should be possible to estimate the value of the trees as timber.

The surveyor and landowner discuss various methods which might be used to pick the 50 trees. They come up with the following ideas:

(a) Drop a pin on the map and take the tree nearest to the point of the pin.

(b) Use a random number generator on a calculator to give 50 numbers between 1 and 200 and select these trees.

(c) Take every 4th tree using the numbers in order.

(d) Divide the area into squares and take the same number of trees in each square.

(e) Count the total number of oaks and divide by 4. Choose that number of oaks at random. Similarly with each of the other varieties.

Activity 6

As a group get everyone to try one of the methods (a) to (e) or one of their own choice. Shade on a copy of the map of the woods the trees you would sample.

Now using the information on the data worksheet on the previous page, find for each method:

(a) the proportion of oaks in your sample.

(b) the average girth of trees.

(c) the average age of the trees.

(d) the proportion of diseased trees.

(e) the tallest tree.

(f) the total value of the woods.

The data columns on the data worksheet show

girth in metres
age in years
disease: 0 - clear, 1 - diseased
height (approx) in metres
value in £.

Using all 200 trees the values are:

(a) 18% (b) 2.15 m (c) 58 years

(d) 16% (e) 9 m (f) £12 925

Compare the results of each method with the overall results. What problems occurred in using the various methods in practice?

The main methods used for sampling in practice are as follows:

(a) **Random** - to be truly random each individual must have an equal chance of being chosen. Dropping a pin on the map is not truly random in this case as it is more likely to select the larger trees. This method is often used for selecting people from Electoral Registers. If the researcher is calling at people's houses the system must be rigidly adhered to (i.e. call back if people are out). It does not necessarily ensure a representative sample.

(b) **Systematic** - taking items at regular intervals e.g. every 4th tree. Although this does not necessarily ensure a representative sample it should be better than random sampling. Again the system must be rigidly adhered to. This method is often used when sampling goods on a production line.

(c) **Stratified** - this is used to ensure that the sample is representative and that it has the same proportions as the population, e.g. ensuring that the sample of trees has the right proportion of each variety. To do this you would need first of all to divide the whole of the population into appropriate categories. This can be very difficult in practice. What is commonly used in street surveys is a **quota** sampling method where interviewers are simply asked to interview a certain proportion of each type, e.g. age, and these can be chosen at random. A common division used is social class. This is defined by the type of job done. The table opposite gives the approximate divisions of social class currently in use.

(d) **Purposive** - in some cases a deliberately biased sample is taken for a particular purpose. If, for example, you wished to test the popularity of a new teenage magazine you would not ask senior citizens. You would, however, ensure the correct proportion of male/female in relation to overall readership.

(e) **Cluster** - sometimes there is a natural sub-grouping of the population - for example, parliamentary constituencies. In this case, you first choose a random sample of clusters and then a sample inside each one. This method can be far less costly than taking a random sample from the whole population.

Composition of Social Classes

Social Class	Main Occupations %
I Professional	Men: engineers and scientists (47.6), accountants (9.2), surveyors (8.5), doctors (5.0), architects (4.7) Women: company secretaries (23.6), doctors (12.7), engineers and scientists (9.5), pharmacists (5.4), clergy and members of religious orders (5.2)
II Intermediate	Men: managers (28.9), proprietors and managers, sales (17.8), teachers (10.6), technicians (9.6), farmers (9.2). Women: teachers (26.7), nurses (24.5), proprietors and managers, sales (16.6), technicians (4.7), managers (4.7).
IIIN Skilled non-manual	Men: clerks, cashiers (51.3), salesmen (20.5), shop assistants (10.6), draughtsmen (8.0), policemen (6.3). Women: clerks, cashiers (46.3), shop assistants (23.7), typists (23.7), office machine operators (4.4).
IIIM Skilled manual	Men: lorry drivers (10.5), lifters (10.5), carpenters (7.2), electricians (5.0), bricklayers (4.9). Women: hairdressers (15.1), cooks (14.1), skilled textile workers (11.7), dressmakers (10.7), printing workers (7.4).
IV Partly skilled	Men: warehousemen (14.4), construction workers (8.8), agricultural workers (8.4), machine tool operators (8.7), metal makers (6.1). Women: maids (18.4), canteen assistants (12.7), partly skilled textile workers (12.6), packers (9.3), telephone operators (4.2).
V Unskilled	Men: labourers (82.6), office cleaners (5.8). Women: office cleaners (64.2), labourers (19.6), kitchen hands (15.1).

Use of random digit tables

For method (b), you could use the random digit table given in the Appendix. Starting arbitrarily on row 10, combining three digits together gives numbers from 000 to 999. Only use numbers in the region 001 to 200; the start of the sequence is:

572 178 878 377 127 957 830 066 ... ...

178 ↑ accept, 127 ↑ accept, 066 ↑ accept

(You normally ignore any repeats if they exist.)

You can attempt to find a random sample more quickly by dividing each three-digit number by 200 and taking the remainder. This would give:

172 178 078 177 127 157 030 066 ...

Would this sample be truly random?

Unfortunately not quite unless 000 is taken as 200, or you take 'the remainder on division by 200 of the three-digit number plus 1'.

Activity 7

Suppose your population is numbered 000 to 299. Use the random digit sheet by taking consecutive three digits. Taking the remainder after division by 300 does **not** give a random sample. Why not?

Activity 8

The map opposite shows a small village of 150 houses (including Ash Farm, Rose Cottage, The Blake Arms and the Shop). The village is due to be redeveloped and the Parish Council wishes to know which of three types of development the village would prefer (these are referred to as C - community centre, H - housing estate, L - large supermarket).

You are asked to undertake a survey of views by sampling 20% of the houses. Use

(a) a systematic sample (b) a random sample

to survey opinion. The views of all the householders are given in the table following. Compare your answers from (a) and (b) with the views of the complete population.

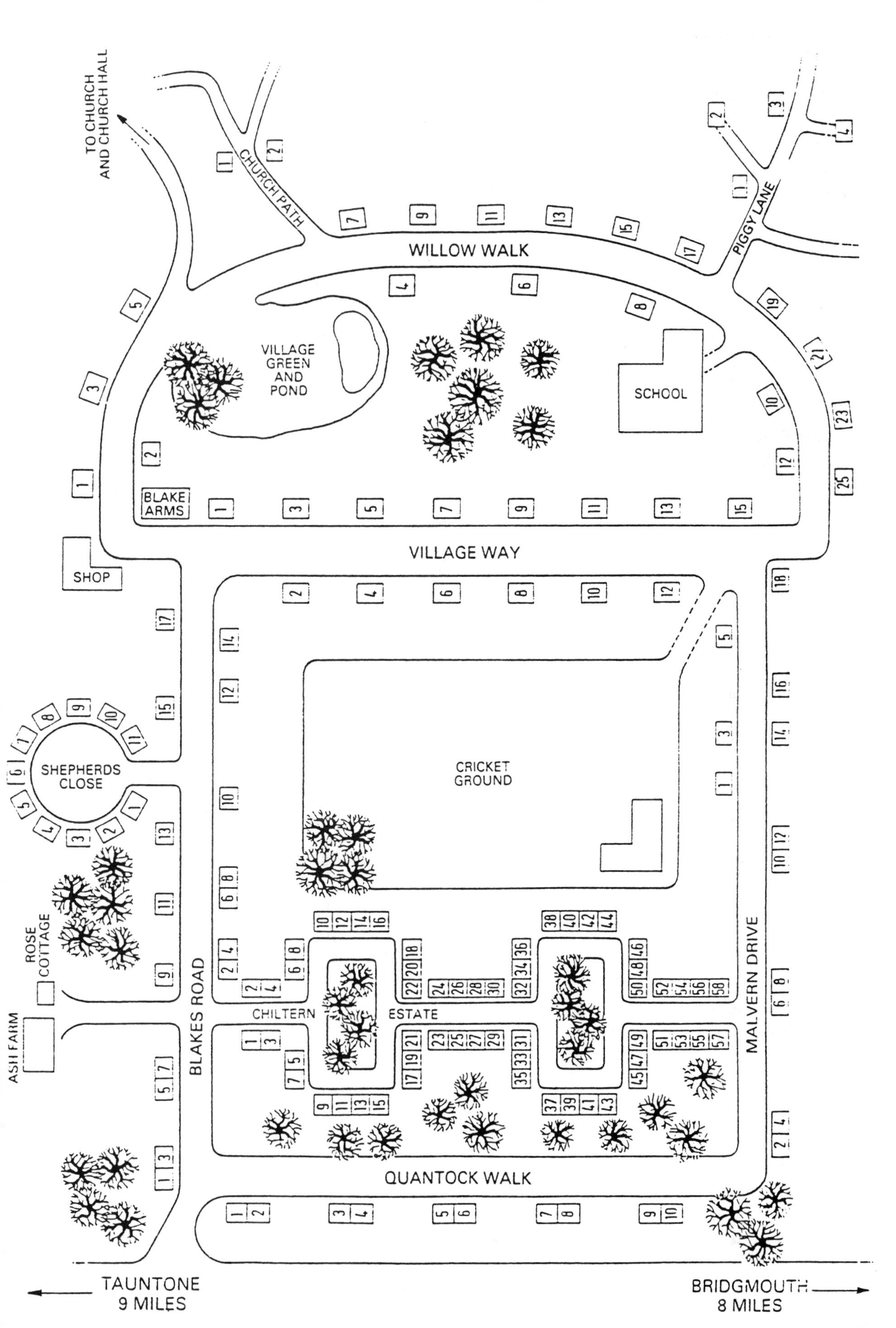

TO CHURCH AND CHURCH HALL
CHURCH PATH
WILLOW WALK
PIGGY LANE
VILLAGE GREEN AND POND
SCHOOL
BLAKE ARMS
VILLAGE WAY
SHOP
CRICKET GROUND
SHEPHERDS CLOSE
ROSE COTTAGE
ASH FARM
BLAKES ROAD
CHILTERN
ESTATE
MALVERN DRIVE
QUANTOCK WALK
TAUNTONE 9 MILES
BRIDGMOUTH 8 MILES

Road	House name or number	Preference
	Rose Cottage	H
	Ash Farm	C
	Blake Arms	H
	Shop	H
Blakes Rd	1	L
	2	L
	3	H
	4	L
	5	H
	6	L
	7	L
	8	C
	9	H
	10	H
	11	C
	12	C
	13	H
	14	L
	15	H
	17	C
Chiltern Estate	1	L
	2	L
	3	H
	4	H
	5	L
	6	L
	7	L
	8	C
	9	L
	10	L
	11	C
	12	H
	13	L
	14	L
	15	L
	16	H
	17	C
	18	L
	19	L
	20	L
	21	L
	22	H
	23	H
	24	H
	25	L
	26	L
	27	C
	28	C
	29	L
	30	L
	31	H
	32	H
	33	C
	34	H
	35	C
	36	C
	37	C
	38	L
	39	L
	40	H
	41	H
	42	L
	43	L
	44	L
	45	C
	46	C
	47	L
	48	L
	49	L
	50	H
	51	H
	52	H
	53	L
	54	H
	55	C
	56	L
	57	H
	58	H

Road	House name or number	Preference
Church Path	1	C
	2	C
Malvern Drive	1	L
	2	L
	3	L
	4	C
	5	H
	6	L
	8	H
	10	C
	12	H
	14	L
	16	H
	18	L
Piggy Lane	1	L
	2	L
	3	H
	4	H
Quantock Walk	1	L
	2	L
	3	C
	4	C
	5	C
	6	H
	7	L
	8	L
	9	L
	10	H
Shepherds Close	1	H
	2	H
	3	C
	4	C
	5	C
	6	H
	7	H
	8	H
	9	L
	10	H
	11	H
Village Way	1	L
	2	H
	3	L
	4	H
	5	C
	6	C
	7	C
	8	L
	9	H
	10	C
	11	H
	12	H
	13	C
	15	C
Willow Walk	1	C
	2	C
	3	C
	4	C
	5	H
	6	H
	7	C
	8	C
	9	H
	10	H
	11	H
	12	H
	13	L
	15	L
	17	H
	19	L
	21	H
	23	H
	25	C

KEY

L – large supermarket
H – housing estate
C – community centre

2.4 Miscellaneous Exercises

1. Pupils in a statistics class want to choose a sample of 100 from a school where the numbers of pupils in each year are shown below.

Year	1	2	3	4	5	6
No. of pupils	290	285	310	175	92	48

(a) Explain how this sample could be obtained by picking a random sample.

(b) If a stratified random sample is chosen, explain how this could be done and how many pupils from each year group are to be chosen for the sample.

2. A factory has 500 employees, each one having a 'works number'. For the purposes of a survey a sample of 25 is picked from the work-force.

Explain

(a) how a systematic sample of 25 could be chosen;

(b) how a random sample, using random numbers, could be chosen;

(c) how a random sample could be chosen, without the use of random numbers.

3. Following a spell of particularly bad weather, an insurance company received 42 claims for storm damage on the same day. Sufficient staff were available to investigate only six of these claims. The others would be paid in full without investigation. The claims were numbered 00 to 41 and the following suggestions were made as to the method used to select the six. In each case six different claims are required, so any repeats would be ignored.

Method 1	Choose the six largest claims
Method 2	Select two-digit random numbers, ignoring any greater than 41. When six have been obtained choose the corresponding claims.
Method 3	Select two digit random numbers. Divide each one by 42, take the remainder and choose the corresponding claims (eg if 44 is selected claim number 02 would be chosen).
Method 4	As 3, but when selecting the random numbers ignore 84 and over.
Method 5	Select a single digit at random, ignoring 7 and over. Choose this and every seventh claim thereafter (e.g. if 3 is selected, choose claims numbered 03, 10, 17, 24, 31 and 38).

Comment on each of the methods, including an explanation of whether it would yield a random sample or not.

4. In a small village, the population is divided by age groups as shown in the table.

Age (years)	0-4	5-14	15-44	45-64	65+
No. of people	14	41	50	70	14

It is proposed to choose a stratified random sample of 40 from the village. Explain how this should be done and calculate how many people should be picked from each age range.

5. Explain briefly what is meant by a random sample. State an advantage of using random, rather than non-random, sampling methods.

Explain the difference between a stratified random sample and a quota sample, and state one advantage of the latter as compared with the former.

An area health authority decides to undertake a survey, using a questionnaire, to determine the proportion of adults who are in favour of local hospitals becoming self-governing trusts. The survey will also investigate patients' attitudes to the treatment presently provided by the hospitals, and aims to collect information from at least 500 adults.

Three possible methods of obtaining the required information are considered.

Method A Choose 1000 adult patients at random from the area's hospitals' records. Arrange for interviewers to visit these patients and ask for the questionnaire to be completed there and then.

Method B Choose names at random from the area's telephone directories. Contact the individuals so chosen, by telephone, and ask if they are willing to answer the questionnaire over the telephone. Continue until enough individuals have agreed to take part.

Method C Choose 2000 names from the area's electoral registers. Send out the questionnaire, by post, to the selected individuals with prepaid envelopes for the questionnaires' return.

(a) Comment critically on the suitability of each of these three methods.

(b) Outline the method you would advise for collecting the required information.

3 DESCRIPTIVE STATISTICS

Objectives

After studying this chapter you should

- understand various techniques for presentation of data;
- be able to use frequency diagrams and scatter diagrams;
- be able to find mean, mode, median, quartiles and standard deviation.

3.0 Introduction

Before looking at all the different techniques it is necessary to consider what the **purpose** of your work is. The data you collected might have been wanted by a researcher wishing to know how healthy teenagers were in different parts of the country. The final result would probably be a written report or perhaps a TV documentary. A straightforward list of all the results could be presented but, particularly if there were a lot of results, this would not be very helpful and would be extremely boring.

The purpose of any statistical analysis is therefore to simplify large amounts of data, find any key facts and present the information in an interesting and easily understandable way. This generally follows three stages:

- sorting and grouping;
- illustration;
- summary statistics.

3.1 Sorting and grouping

The following table shows in the last two columns the average house prices for different regions in the UK in 1988 and 1989.

Clearly prices have increased but has the pattern of differences between areas altered?

	% dwellings owner occupied		**Average dwelling price (£)**	
	1988 (end)	**1989 (end)**	**1988**	**1989**
United Kingdom	65	67	49 500	54 846
North	58	59	30 200	37 374
Yorks. and Humbs.	64	66	32 700	41 817
East Midlands	69	70	40 500	49 421
East Anglia	68	70	57 300	64 610
South East	68	69	74 000	81 635
South West	72	73	58 500	67 004
West Midlands	66	67	41 700	49 815
North West	67	68	34 000	42 126

(Source: United Kingdom in Figures - Central Statistical Office)

One simple way you could look at the data is to place them all in order, e.g. for 1988 prices:

North	30 200
Yorks & Humbs.	32 700
North West	34 000
East Midlands	40 500
West Midlands	41 700
East Anglia	57 300
South West	58 500
South East	74 000

Even a simple exercise such as this shows clearly the range of values and any natural groups in the data and allows you to make judgements as to a typical house price.

However, with larger quantities of data, putting into order is both tedious and not very helpful. The most commonly used method of storing large quantities of data is a **frequency** table. With qualitative or discrete quantitative data this is simply a record of how many of each type were present. The following frequency table shows the frequency with which **other types of vehicles** were involved in cycling accidents:

	Number	%
Motor Cycle	96	2.5
Motor Car	2039	52.3
Van	168	4.3
Goods Vehicle	126	3.2
Coach	49	1.3
Pedestrian	226	5.8
Dog	120	3.1
Cyclist	218	5.6
None - defective road surface	266	6.8
None - weather conditions	129	3.3
None - mechanical failure	65	1.7
Other	399	10.2
Total	3901	100.1

(Source: Cycling Accidents - Cyclists' Touring Club)

With continuous data and with discrete data covering a wide range it is more useful to put the data into groups. For example, take the share prices in the information in the last section. This could be recorded as shown below:

Share Price (p)	Frequency
1 - 200	....
201 - 400	....
401 - 600	....
601 - 800	....
801 - 1000	....
1001 ormore	....
Total	

Note the following points:

- Group limits do not overlap and are given to the same degree of accuracy as the data is recorded.
- Whilst there is no absolute rule, neither too many nor too few groups should be used. A good rule is to look at the range of values, taking care with extremes, and divide into about six groups.
- If uneven group sizes are used this can cause problems later on. The only usual exception is that 'open ended' groups are often used at the ends of the range.

- The class boundaries are the absolute extreme values that could be rounded into that group, e.g. the upper class boundary of the first group is 200.5 (really 200.4999.....).

Stem and leaf diagrams

A new form of frequency table has become widely used in recent years. The **stem and leaf** diagram has all the advantages of a frequency table yet still records the values to full accuracy.

As an example, consider the following data which give the marks gained by 15 pupils in a Biology test (out of a total of 50 marks):

27, 36, 24, 17, 35, 18, 23, 25, 34, 25, 41, 18, 22, 24, 42

The stem and leaf diagram is determined by first recording the marks with the 'tens' as the **stem** and the 'units' as the **leaf**.

This is shown opposite.

Stem	Leaf
0	
1	7 8 8
2	7 4 3 5 5 2 4
3	6 5 4
4	1 2

The leaf part is then reordered to give a final diagram as shown. This gives, at a glance, both an impression of the spread of these numbers and an indication of the average.

Stem	Leaf
0	
1	7 8 8
2	2 3 4 4 5 5 7
3	4 5 6
4	1 2

Example

Form a stem and leaf diagram for the following data:

21, 7, 9, 22, 17, 15, 31, 5, 17, 22, 19, 18, 23,

10, 17, 18, 21, 5, 9, 16, 22, 17, 19, 21, 20.

Solution

As before, you form a stem and leaf, recording the numbers in the leaf to give the diagram opposite.

Stem	Leaf
0	5 5 7 9 9
1	0 5 6 7 7 7 7 8 8 9 9
2	0 1 1 1 2 2 2 3
3	1

Exercise 3A

1. For each of the measurements you made at the start of Chapter 2 compile a suitable frequency table, or if appropriate a stem and leaf diagram.
2. The table below shows details of the size of training schemes and the number of places on the schemes. Notice that the table has used uneven group sizes. Can you suggest why this has been done?

Size of Training Schemes

Number of approved places	Number of schemes	Percentage of all schemes
1– 20	2167	51.4
21– 50	855	20.3
51– 100	581	13.8
101– 500	560	13.3
501– 1000	41	1.0
over 1000	14	0.3
	4218	**100.0**

(Source: August 1985 Employment Gazette)

3. The table below shows the ages of registered drug addicts in the period 1971 -1976. What conclusions can you draw from this about the relative ages of drug users during this period?

Dangerous drugs: registered addicts United Kingdom

	1971	1972	1973	1974	1975	1976
Males	1133	1194	1369	1459	1438	1389
Females	416	421	446	512	515	492
Age distribution:						
Under 20 years	118	96	84	64	39	18
20 and under 25	772	727	750	692	562	411
25 and under 30	288	376	530	684	754	810
30 and under 35	112	117	134	163	219	247
35 and under 50	112	118	136	163	169	189
50 and over	177	165	180	197	193	188
Age not stated	20	16	1	8	17	18

3.2 Illustrating data - bar charts

In the last question of the previous exercise you would have to look at the different figures and make size comparisons to interpret the data; e.g. in 1976 there were twice as many in the 25-30 age group as were in the 20-25 age group. Using diagrams can often show the facts far more clearly and bring out many important points.

The most commonly used diagrams are the various forms of **bar chart**. A true bar chart is strictly speaking only used with qualitative data, as shown opposite.

Child pedestrians killed in Europe: deaths per million population

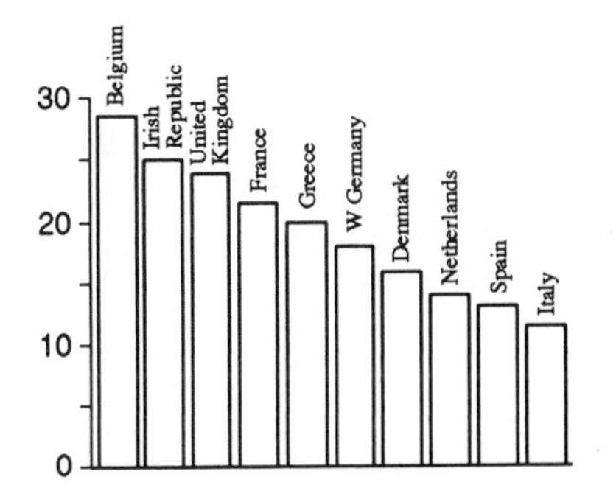

Note that there is no scale on the horizontal axis and gaps are left between bars.

With quantitative discrete data a frequency diagram is commonly used. In a school survey on the number of passengers in cars driving into Norwich in the rush hour the following results were obtained.

No. of passengers	Frequency
0	13
1	25
2	12
3	6
4	1

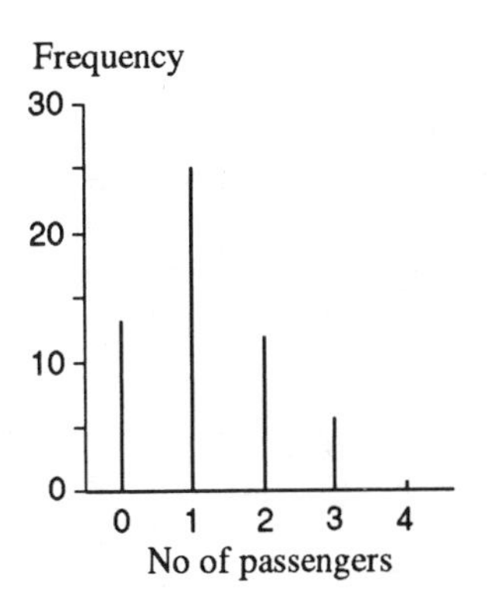

Strips are used rather than bars to emphasise discreteness. In practice, however, many people use a bar as this can be made more decorative. It is again usual to keep the bars separate to indicate that the scale is not continuous.

Composite bar charts

Composite bar charts are often used to show sets of comparable information side by side, as shown opposite.

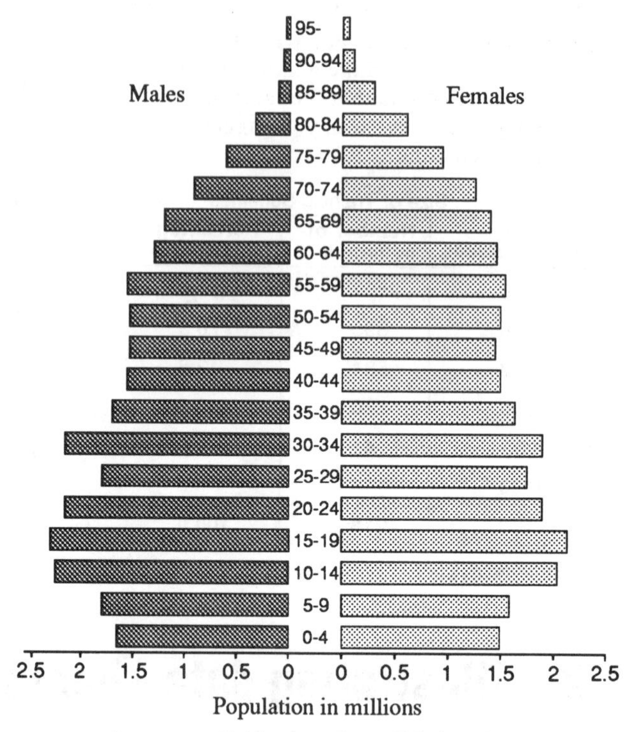

Age group distribution, Great Britain, 1981

There are alternative ways this could have been shown, as illustrated opposite and below.

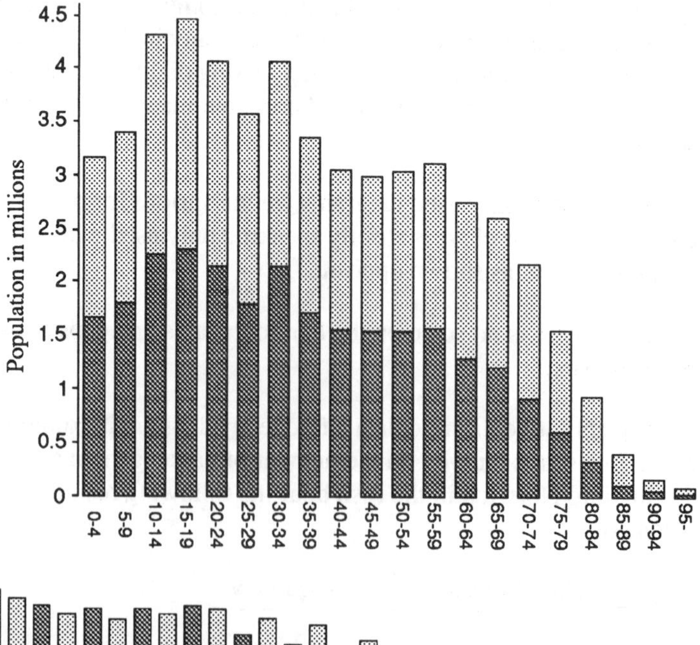

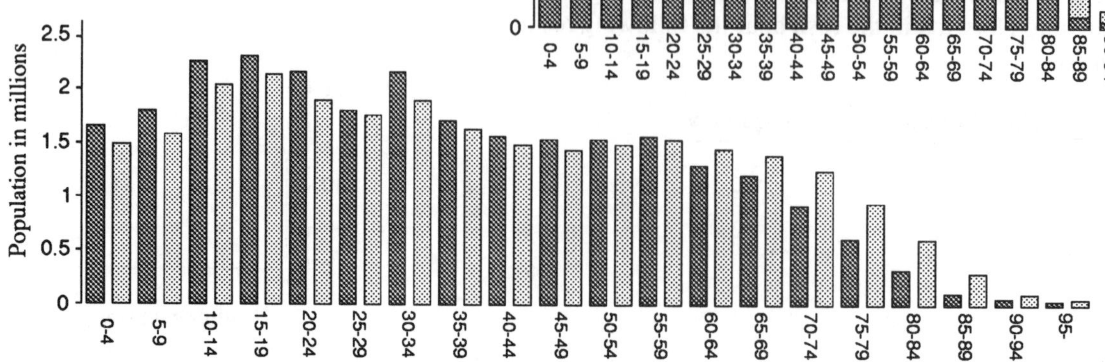

Activity 1 Interpreting the graph

Working in groups, consider these questions about the previous composite bar charts.

What are the main differences between the age distributions of men and women? Can you explain why there are more people in their 50's than 40's? What are the main advantages and disadvantages of each of the different methods of presenting the data?

Histograms

A **histogram** is generally used to describe a bar chart used with continuous data.

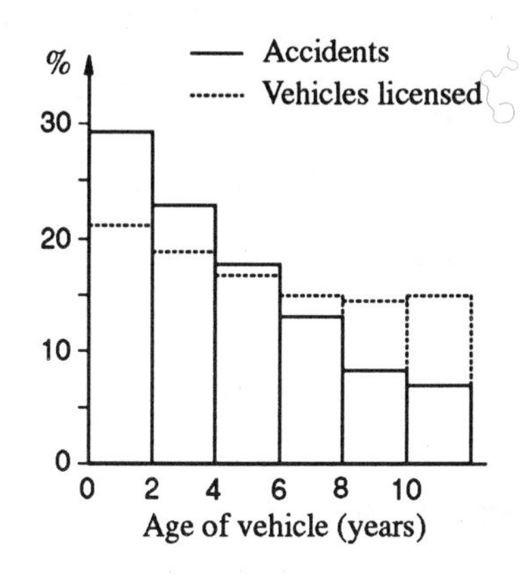

Note that the horizontal axis is a proper numerical scale and that no gaps are drawn between bars. Bars are technically speaking drawn up to the class boundaries though in practice this can be hard to show on a graph. Care must be taken however if there are uneven group sizes. For example the following table shows the percentages of cyclists divided into different age groups and sexes.

Number of years cycling	Age 0-16	16-25	25+	Sex Male	Female
0-1	6%	4%	1%	2%	3%
1-2	18%	8%	3%	4%	8%
2-5	35%	25%	10%	12%	21%
5-10	31%	29%	9%	13%	15%
10-14	9%	33%	77%	69%	52%

(Source: Cycling Accidents - Cyclist's Touring Club.)

If you use the pure frequency values from the table to draw a histogram showing the percentages of children aged 0-16 who have been cycling for different numbers of years, you get the diagram opposite. This, though, is incorrect .

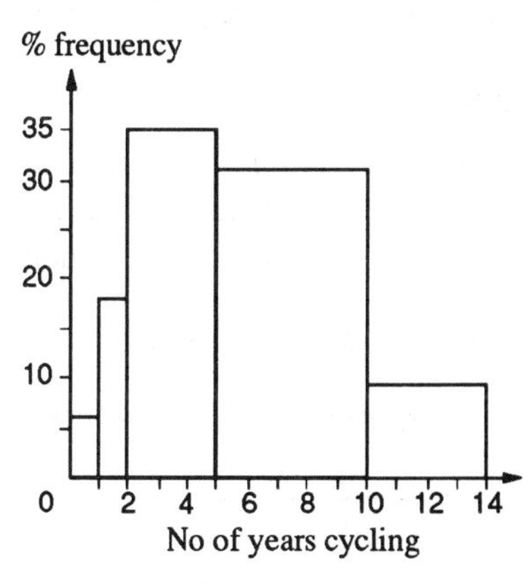

The fact that the groups are of different widths makes it appear that children are more likely to have been cycling for longer periods. This is because our eyes look at the proportion of the **areas**. To overcome this you need to consider a standard unit, in this case a year. The first two percentage frequencies would be the same, but the next would be 35/3 = 11.7% as it covers a three year period. This is called the **frequency density**; that is, the frequency divided by the class width. Similarly, dividing by 5 and 4 gives the heights for the remaining groups. The correct histogram is shown opposite.

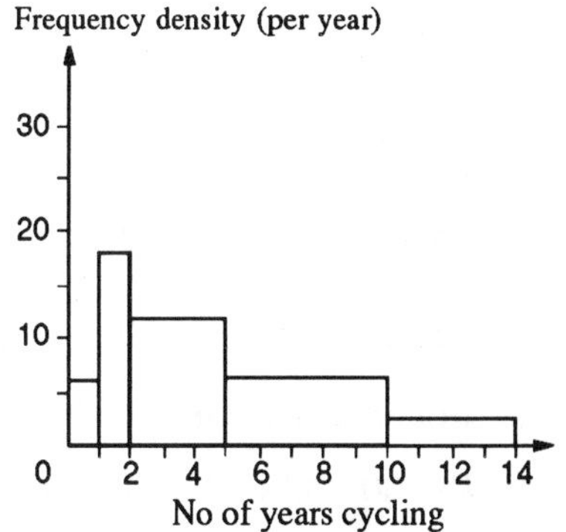

Note the labelling of the vertical scale.

Example

The table shows the distribution of interest paid to investors in a particular year.

Interest (£)	25-	30-	40-	60-	80-	110-
Frequency	18	55	140	124	96	0

Draw a histogram to illustrate the data.

Solution

Interest	Class widths	Frequency	Frequency density
25-	5	18	3.6
30-	10	55	5.5
40-	20	140	7.0
60-	20	124	6.2
80-	30	96	3.2

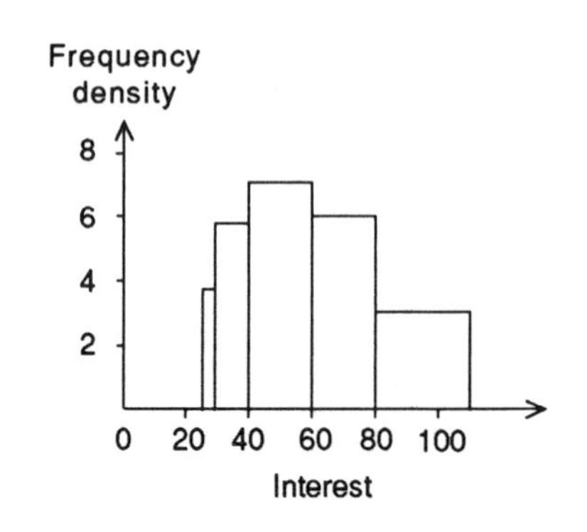

Example

The histogram opposite shows the distribution of distances in a throwing competition.

(a) How many competitors threw less than 40 metres?

(b) How many competitors were there in the competition?

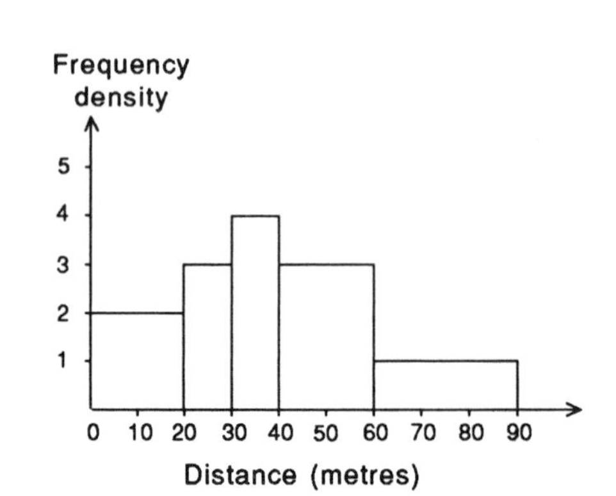

Solution

Using the formula

$$\text{class width} \times \text{frequency density} = \text{frequency}$$

gives the following table.

Interval	Class width	Frequency density	Actual frequency
0-20	20	2	$2\times 20=40$
20-30	10	3	$3\times 10=30$
30-40	10	4	$4\times 10=40$
40-60	20	3	$3\times 20=60$
60-90	30	1	$1\times 30=30$

(a) $40+30+40=110$

(b) $40+30+40+60+30=200$

There are a number of common shapes which appear in histograms and these are given names:

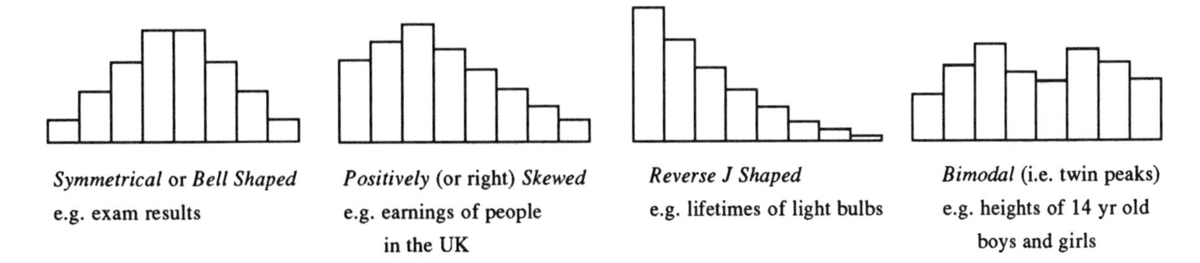

Symmetrical or *Bell Shaped* e.g. exam results

Positively (or right) *Skewed* e.g. earnings of people in the UK

Reverse J Shaped e.g. lifetimes of light bulbs

Bimodal (i.e. twin peaks) e.g. heights of 14 yr old boys and girls

When a histogram is drawn with continuous data it appears that there are shifts in frequency at each class boundary. This is clearly not true and to show this you can often draw a line joining the middles of the tops of the bars, either as a series of straight lines to form a **frequency polygon**, or more realistically with a curve to form a **frequency curve**. These also show the shape of the distribution more clearly.

Exercise 3B

1. Draw appropriate bar charts for the data you collected at the start of Chapter 2.
2. Use the information on the ages of sentenced prisoners in the table opposite to draw a composite bar chart. Ignore the uneven group sizes.

   Explain why you have used the particular type of diagram you have.

Age and sex of prisoners, England and Wales 1981

Age	Men	Women
14-16	1637	129
17-20	9268	238
21-24	7255	235
25-29	5847	188
30-39	7093	236
40-49	3059	132
50-59	1128	35
60 and over	262	7

3. The information below and opposite relates to people taking out mortgages. Draw an appropriate bar chart for the All buyers information in each case.

By type of dwelling (%)

Type	All buyers
Bungalow	10
Detached house	19
Semi-detached house	31
Terraced house	31
Purpose built flat	7
Converted flat	3

By age of borrowers (%)

Age	All buyers
Under 25	22
25-29	26
30-34	21
35-44	20
45-54	8
55 & over	3

By mortagage amounts(%)

Amount	All buyers
Under £8000	16
£ 8000 - £ 9999	10
£10000 - £11999	16
£12000 - £13999	17
£14000 - £15999	17
£16000 & over	24

4. 100 people were asked to record how many television programmes they watched in a week. The results are shown opposite.

   Draw a histogram to illustrate the data.

No. of programmes	0-	10-	18-	30-	35-	45-	50-	60-
No. of viewers	3	16	36	21	12	9	3	0

5. 68 smokers were asked to record their consumption of cigarettes each day for several weeks. The table shown opposite is based on the information obtained.

   Illustrate these data by means of a histogram.

Average no. of cigarettes smoked per day	0-	8-	12-	16-	24-	28-	34-50
No. of smokers	4	6	12	28	8	6	4

3.3 Illustrating data - pie charts

Another commonly used form of diagram is the **pie chart**. This is particularly useful in showing how a total amount is divided into constituent parts. An example is shown opposite.

QUESTION
Do you think girls are better off going to single sex or mixed schools?

QUESTION
Do you think boys are better off going to single sex or mixed schools?

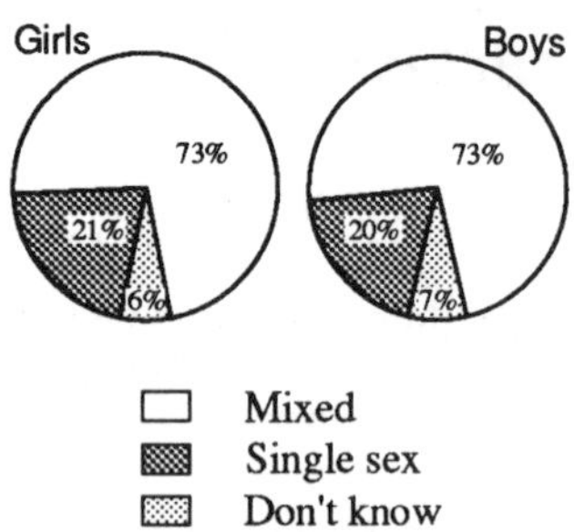

To construct a pie chart it is usually easiest to calculate percentage frequencies. Look at the contents list for the packet of 'healthy' crisps:

Nutrient	Per 100 g
Protein	6.1 g
Fat	34.2 g
Carbohydrates	48.1 g
Dietary Fibre	11.6 g

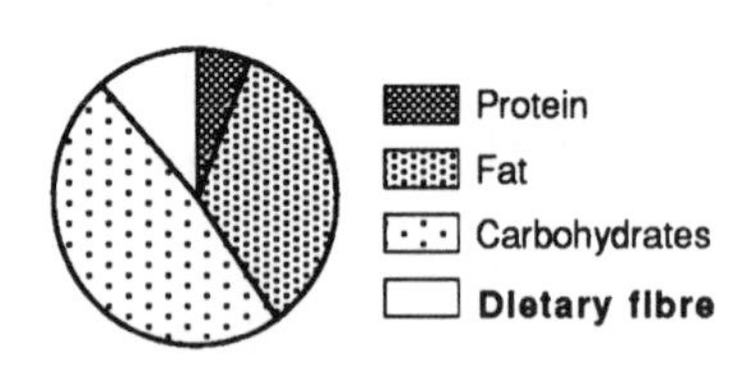

There are now percentage pie chart scales which can be used to draw the charts directly. Using a traditional protractor method you need to find 6.1% of 360° etc. This gives the pie chart shown above.

When two sets of information with different totals need to be shown, the comparative pie charts are made with sizes proportional to the totals. However, as was discussed with histograms, it is the relative area that the mind uses to make comparisons. The radii therefore have to be in proportion to the **square root** of the total proportion. For example, in the graph opposite the pie charts are drawn in proportion to the 'average total expenditure' i.e. 59.93/28.52 = 2.10.

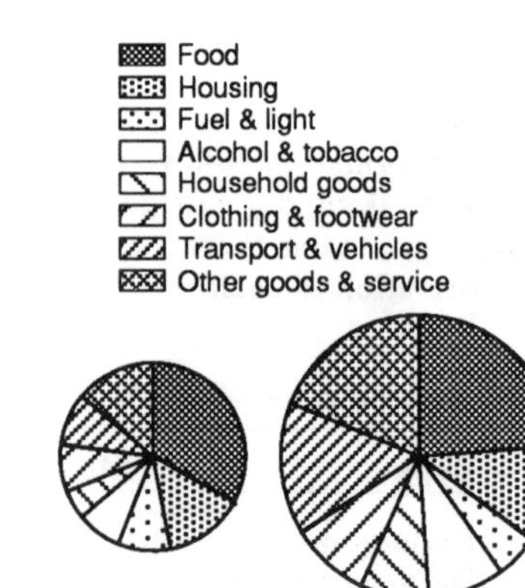

Low income households
Average total expenditure
£28.52 per week

Other households
Average total expenditure
£59.93 per week

The radii are therefore in the proportion $\sqrt{2.10} \approx 1.45$. Smaller radius $= 1.7$ cm, then the larger radius $= 1.7 \times 1.45 = 2.5$ cm.

In general, when the total data in the two cases to be illustrated are given by A_1 and A_2, then the formula for the corresponding radii is given by

$$\frac{A_1}{A_2} = \frac{\pi r_1^2}{\pi r_2^2} = \left(\frac{r_1}{r_2}\right)^2$$

Alternatively,

$$\frac{r_1}{r_2} = \sqrt{\frac{A_1}{A_2}}$$

Exercise 3C

1. Draw pie charts for hair colour and eye colour from the results of your survey at the start of Chapter 2.

2. During the 1983 General Elections the % votes gained by each party and the actual number of seats gained by each party are shown opposite.

   (a) Draw separate pie charts, using the same radius, for votes and seats won.

   (b) Calculate the number of seats that would have been gained if seats were allocated in proportion to the % votes gained. Show this and the actual seats gained on a composite bar chart.

   (c) Show how this information could be used to argue the case in favour of proportional representation.

	Conservative	Labour
% Votes	43.5	28.3
Seats won	397	209

	Liberal/Democrats	Other
% Votes	26.0	2.2
Seats won	23	21

3. According to a report showing the differences in diet between the richest and poorest in the UK the figures opposite were given for the consumption of staple foods (ounces per person per week).

   Draw comparative pie charts for this information. What differences in dietary pattern does this information show?

	Poorest 10%	Richest 10%
White bread	26	12.3
Sugar	11.5	8
Potatoes	48.3	33.4
Fruit	13	25.3
Vegetables	21.5	30.7
Brown bread	5.2	8

3.4 Illustrating data - line graphs and scattergrams

Where there is a need to relate one variable to another a different form of diagram is required. When a link between two different quantities is being examined a **scattergram** is used. Each pair of values is shown as a point on a graph, as shown opposite.

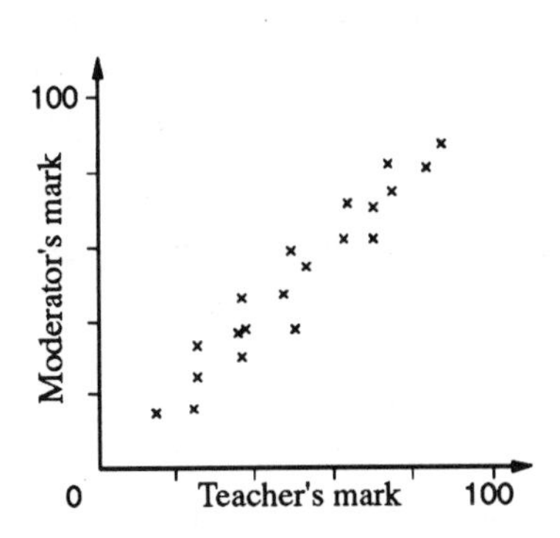

In other cases where the scale on the x-axis shows a systematic change in a particular time period, a line graph can be used as shown in the graph opposite.

The effect of a popular television programme on electricity demand is shown in this curve, which shows typical demand peaks. Peaks A and E coincide with the start and finish of the programme; peaks B, C and D coincide with commercial breaks.

Care needs to be taken over vertical scales. In the graph opposite it appears that the value of the peseta has varied dramatically in relation to the pound. However, looking at the scale shows that this has at most varied by 20 pesetas ($\pm 5\%$). To start the scale at 0 would clearly be unreasonable so it is usual to use a zig-zag line at the base of a scale to show that part of the scale has been left out.

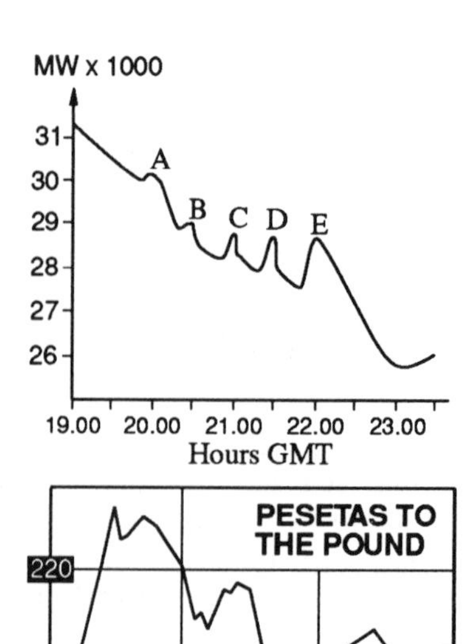

Exercise 3D

1. By drawing scattergrams of your data from Activity 1 at the start of Chapter 2 examine the following statements:

   (a) Taller people tend to have faster pulses.

   (b) People with faster pulses tend to have quicker reaction times.

   (c) High blood pressure is more common in heavier people.

2. The next page shows details of statistics published by Devon County Council on road accidents in 1991. Use this information to write a newspaper report on accidents in the county that year. Include in your report any of the tables and diagrams shown or any of your own which you think would be suitable in an article aimed at the general public.

3.5 Using computer software

There are many packages available on the market which are able to do all or most of the work covered here. These fall into two main categories:

(a) Specific statistical software where a program handles a particular technique and data are fed in directly.

(b) Spreadsheet packages, where data are stored in a matrix of rows and columns; a series of instructions can then carry out any technique which the particular package is able to do.

In the commercial/research world very little work is now carried out by hand; the large quantities of data would make this very difficult.

Activity 2

If you have access to a computer, find out what software you have available and use this to produce tables and diagrams for the data you have collected.

How many?

Reported injury accidents have decreased by 11% compared with last year. Traffic flows also show a small decrease in numbers in urban areas.

Accidents by year and severity

Year	Fatal	Serious	Slight	Total injury accidents
82	91	1 521	2 680	4 292
83	87	1 453	2 808	4 348
84	78	1 486	2 868	4 432
85	65	1 432	3 003	4 500
86	78	1 424	2 950	4 452
87	81	1 243	2 891	4 215
88	74	1 188	3 056	4 318
89	80	1 120	3 199	4 399
90	67	1 048	3 124	4 239
91	76	866	2 814	3 756

Target reduction

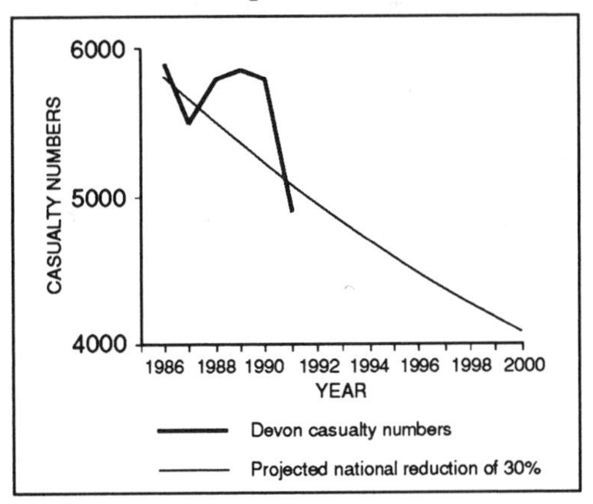

The government has set a target of 30% reduction in casualties by the year 2000 using a base of an average figure for 1981 - 1985.

Who?

This table shows the number of people killed and injured in 1991.

Casualties by road user type

	1991			
	Fatal	Serious	Slight	Total
Pedestrians	21	216	497	734
Pedal Cyclists	2	69	257	328
Motorcycle Riders	21	234	431	686
Motorcycle Passengers	0	14	50	64
Car Drivers	20	265	1387	1672
Front Seat Car Passengers	7	110	590	707
Rear Seat Car Passengers	6	61	325	392
Public Service Vehicle Passengers	0	4	67	71
Other Drivers	4	26	17	147
Other Passengers	2	14	43	59
Totals	83	1013	3764	4860

Injury accidents by day of week 1991

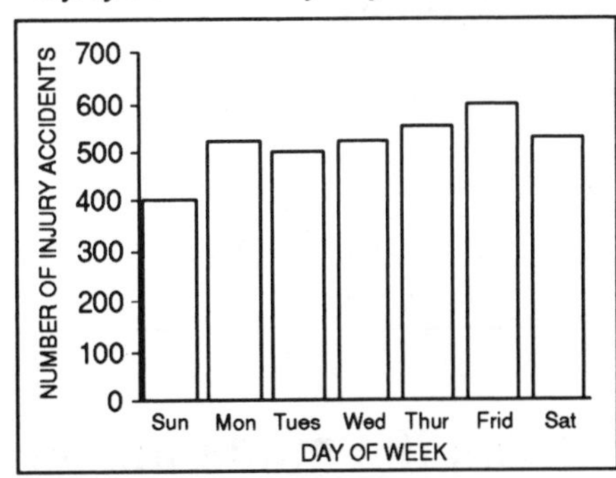

Accident levels are highest towards the end of the week. This reflects the increased traffic on those days during holiday periods as well as weekend 'evenings out' throughout the year.

Accidents involving children

The table shows the number of children killed and injured in Devon for the years 1989 - 1991.

	Age group (years)											
	0 - 4			5 - 9			10 - 15			Total 0 - 15		
	89	90	91	89	90	91	89	90	91	89	90	91
Pedestrians	41	48	49	96	105	89	139	125	112	276	278	250
Pedal cycles	1	1	2	25	20	27	134	115	105	160	136	134
Car passengers	38	71	38	72	54	49	107	93	88	217	218	175
Others	2	12	4	4	16	5	68	46	18	74	74	27
Totals	82	132	93	197	195	170	448	379	323	727	706	586

Injury accidents by time of day 1991

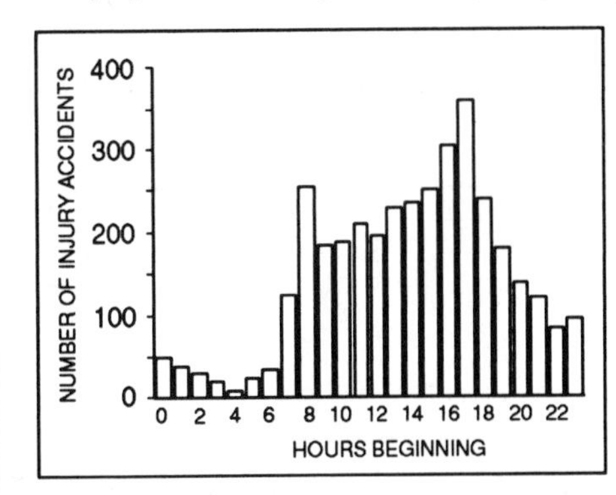

Accidents plotted by hours of day clearly shows the peaks during the rush hours particularly in the evening. Traffic flows decrease during the rest of the evening but the accident levels remain high.

3.6 What is typical?

At the beginning of this chapter the question was posed 'what is the 'normal' blood pressure for someone of your age'. If you did this experiment you will perhaps have a better idea about what kind of value it is likely to be. Another question you might ask is 'Are women's blood pressures higher or lower than men's?'

If you just took the blood pressure of one man and one woman this would be a very poor comparison. What you need, therefore, is a single representative value which can be used to make such comparisons.

Activity 3

Obtain about 30 albums of popular music where the playing time of each track is given. Write down the times in decimal form (most calculators have a button which converts minutes and seconds to decimal form) and the total time of the album. Also write down the number of tracks on the album.

There are two questions that could be asked:

(a) What is a typical track/album length?

(b) What is a typical number of tracks on an album?

Using the mode and median

The easiest measure of the average that could be given is the **mode**. This is defined as the item of data with the **highest frequency**.

Activity 4 Census data

An extract from the 1981 census is shown opposite. What does it show?

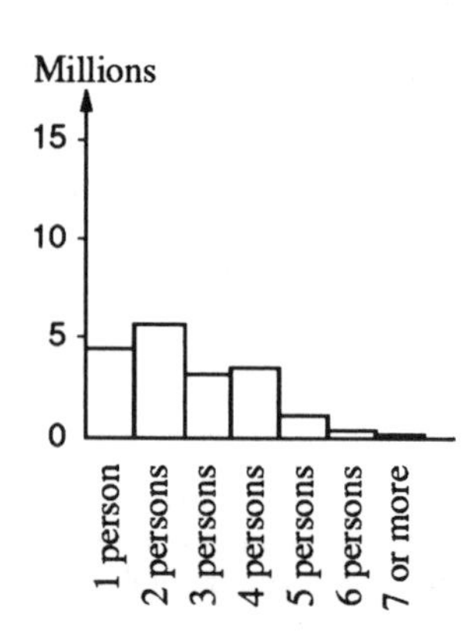

SIZE OF HOUSEHOLDS

The most common size of household in 1981 was two people. There were just under 20 million households in total.
In 4.3% of households in Great Britain there was more than one person per room compared with 7.2% in 1971.

When data are grouped you have to give the **modal group**. In the following example the modal group is 1500 cc - 1750 cc.

Engine size : Private cars involved in accidents

-1000 cc	7.7%
-1250 cc	13.9%
-1500 cc	25.4%
-1750 cc	27.2%
-2000 cc	12.6%
-2500 cc	9.3%
Over 2500 cc	3.9%

(Source - Analysis of accidents - Assn. of British Insurers)

There are, however, problems with using the mode:

(a) The mode may be at one extreme of the data and not be typical of all the data. It would be wrong to say from the data opposite that accidents were typically caused by people who had passed their test in the last year.

(b) There may be no mode or more than one mode (bimodal).

(c) Some people use a method with grouped data to find the mode more precisely within a group. However, the way in which data were grouped can affect in which group the mode lies.

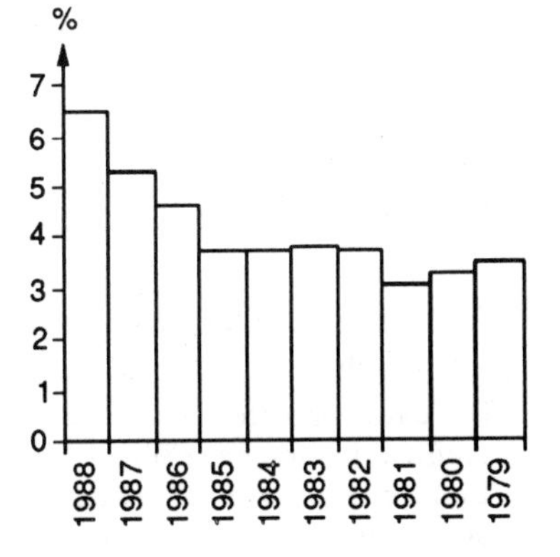

Distribution of accidents in 1989 by year in which driving test was passed.

The mode has some practical uses, particularly with discrete data (e.g. tracks on an album) and you can even use the mode with qualitative data. For example, a manufacturer of dresses wishing to try out a new design in one size only would most likely choose the modal size.

The **median** aims to avoid some of the problems of the mode. It is the value of the **middle item** of data when they are all placed in order. For example, to find the median of a group of seven people's weights in kg: 75.3, 82.1, 64.8, 76.3, 81.8, 90.1, 74.2, you first put them in order and then identify the middle one.

64.8, 74.2, 75.3, 76.3, 81.8, 82.1, 90.1,
↑
median

Example

Find the median mark for the following exam results (out of 20). Compare this to the mode.

2, 3, 7, 8, 8, 8, 9, 10, 10, 11, 12, 12, 14, 14, 16, 17, 17, 19, 19, 20

Solution

There are 20 items of data, so the median is the $\frac{21}{2} = 10\frac{1}{2}$th item;

i.e. you take the average of the 10th and 11th items, giving

$$\text{median} = \frac{11+12}{2} = \frac{23}{2} = 11.5.$$

The mode is 8, since there are three results with this value.

For these data, the median gives a more representative mark than does the mode.

In general, if there are n items of data, the median is the

$$\frac{(n+1)}{2}\text{th item.}$$

Where there is an even number of data the median will be in between two actual values of data, and so the two values are averaged.

Exercise 3E

1. Find the median length of track time for each of your albums.
2. The data opposite show the cost of various medical insurance schemes for people living in London or provincial areas. Find the median cost of insurance for a single person aged 25 in

   (i) London (ii) Provincial areas.

   What is the approximate extra paid by a person living in London?

Company	Maximum benefits yearly per person	Yearly premium for single person (age 25)	
		London rates	Provincial rates
	£	£	£
AMA	40 000	222	153
BCWA	No limit	190	139
BUPA	No limit	316	205
Crown Life	45 000	258	172
Crusader	No limit	279	195
EHAS	No limit	292	236
Health First	No limit	255	166
Holdcare	No limit	180	134
Orion	50 000	182	182
PPP	No limit	288	156
WPA	45 000	271	188

3.7 Grouped data

With grouped data a little more work is required. An example concerning yearly cycling in miles is shown opposite.

The median is the

$$\frac{(8552+1)}{2} = 4276.5\text{th item.}$$

There are two commonly used methods for finding this:

Miles cycled in 1980

Miles	Number	%
0-500	1252	15
500-1000	1428	17
1000-1500	1231	14
1500-2000	1016	12
2000+	3625	42
TOTAL	8552	100

(a) **Linear interpolation**. This assumes an even spread of data within each group.

By adding up the frequencies:

$$1252+1428+1231=3911$$

but $\quad 3911+1016=4927$

You can deduce that the 4276.5 th piece of data is therefore in the 1500–2000 group and in the bottom half.

More precisely this is $4276.5-3911=365.5$ items along that group. Since there are 1016 item in this group you need to go $365.5/1016 = 0.36$ of the way up this group.

This will be

$$1500+(0.360\times 500)=1680.$$

It should be remembered this is only an approximate result and should not be given to excessive accuracy.

(b) **Cumulative frequency curves**. This is a graphical method and therefore of limited accuracy, but assumes a more realistic nonlinear spread in each group. Other information apart from the median can also be obtained from them.

The cumulative frequencies are the frequencies that lie below the upper class boundaries of that group. For example in a large survey on people's weights in kg the following results were obtained:

Weight (kg)	Frequency	Cumulative frequency
< 33	1	1
33.0 - 33.9	0	1
34.0 - 34.9	2	3
35.0 - 35.9	8	11
36.0 - 36.9	19	30
37.0 - 37.9	27	57
38.0 - 38.9	25	82
39.0 - 39.9	14	96
40.0 - 49.9	3	99
≥ 50.0	1	100

For example, the cumulative frequency 30 tells you that 30 people weighed less than 36.95 kg. These are then plotted using the **upper class boundaries** (U.C.B.) on the x-axis.

The median is at the 50.5th item and can be read from the graph. The graph can also be used to answer such questions as, 'How many people weighed 38.5 kg or less?

Note the 'S' shape of the graph, which will occur when the distribution is bell shaped.

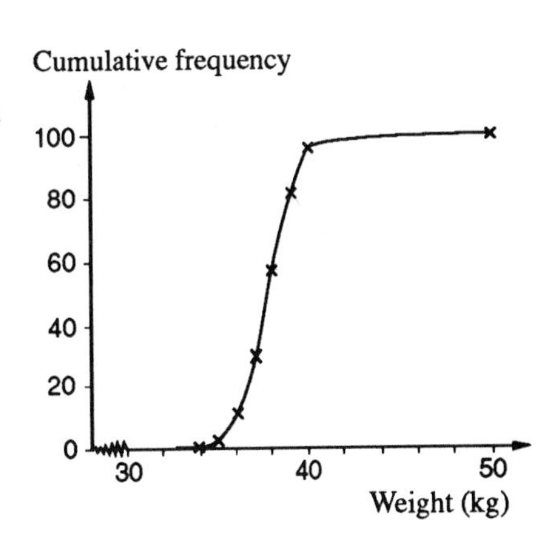

Activity 5

Use the cumulative frequency graph on page 63 to estimate

(a) the percentage of people with weight

(i) less than 38.5 kg,

(ii) greater than 37.5 kg;

(b) the weight which is exceeded by 75% of people.

Exercise 3F

1. Draw up a frequency table of the track times for all the albums in the survey conducted in Activity 3. Draw a cumulative frequency curve of the results and use this to estimate the median playing time.

2. The data below show the monthly rainfall at various weather stations in Norfolk one September. Compile a frequency table and draw a cumulative frequency curve to find the median monthly rainfall.

Station	Rainfall	Station	Rainfall	Station	Rainfall	Station	Rainfall
Acle	91.6	Dunton	67.6	Lingwood	79.2	U.Sheringham	71.4
Ashi	80.8	Edgefield	H108.4	Loddon	74.0	Shotesham	82.0
Ayylebridge	74.8	Fakenham	84.3	Lyng	74.8	Shropham	85.6
Aylsham	91.4	Felmingham	85.9	Marham R.A.F.	59.5	Snettisham	82.3
Barney	82.5	Feltwell	71.6	Morley	78.7	Snoring Little	79.0
Barton	84.7	Foulsham	78.76	Mousehold	74.8	Spixworth	72.0
Bawdeswell	73.2	Framingham C	69.6	Norton Subcourse	69.3	Starston	78.5
Beccles	73.7	Fritton	82.0	Norwich Cemetery	84.8	S.Strawless	77.2
Besthorpe	73.5	Great Fransham	75.5	Nch.G Borrow Road	85.3	Swaffham	87.9
Blakeney	76.1	Gooderstone	75.1	Ormesby	94.7	Syderstone	88.2
Braconash	57.9	Gressehall	71.4	Paston School	81.9	Taverham	83.4
Bradenham	58.4	Heigham WW	87.7	Pulham	68.5	North Thorpe	78.6
Briston	91.5	Hempnall	66.9	Raveningham	44.7	Thurgarton	70.0
Brundall	68.6	Hempstead Holt	105.5	E.Raynham	70.5	Tuddenham E	79.8
Burgh Castle	76.9	Heydon	76.2	S.Raynham	78.1	Tuddenham N	81.5
Burnham Market	63.0	Hickling	63.2	Rougham	72.9	Wacton	61.6
Burnham Thorpe	L42.2	Hindringham	65.8	North Runeton	61.7	North Walsham	75.2
Buxton	85.3	Holme	69.3	Saham Toney	84.3	West Winch	65.9
Carbrooke	93.1	Hopton	84.9	Salle	75.0	Gt. Witchingham	74.7
Clenchwarton	56.0	Horning	87.7	Sandringham	76.5	Wiveton	78.2
Coltishall R.A.F.	87.0	Houghton St. Giles	89.2	Santon Downham	89.4	Wolferton	59.0
Costessey	74.6	Ingham	75.2	Scole	71.3	Wolterton Hall	89.8
North Creake	80.2	High Kelling	93.5	Sedgeford	65.8	Woodrising	82.9
Dereham	85.8	Kerdiston	73.2	Shelfanger	76.6	Wymondham	68.2
Ditchingham	67.6	King's Lynn	63.5	L.Sheringham	72.8	Taverh'm 46-yr av.	53.6
Downham Market	59.7	Kirstead	79.2				

H - highest, L - lowest

(Source : Eastern Daily Press)

3. The distribution of ordinary shares for Cable & Wireless PLC in 1987 is shown opposite. Find the median amount of shares using interpolation. Comment critically on the use of the median as a typical value in this case.

The distribution of ordinary shares at 31 March, 1987	Number of holdings
1 - 250	50 268
251 - 500	69 443
501 - 1 000	25 705
1 001 - 10 000	32 730
10 001 - 100 000	2 086
100 001 - 999 999	669
1 000 000 and over	166
	181 067

(Source: Cable & Wireless PLC - Report 1987)

3.8 Interpreting the mean

One criticism of the median is that it does not look at **all** the data. For example a pupil's marks out of 10 for homework might be:

3, 4, 4, 4, 9, 10, 10.

The pupil might think it unfair that the median mark of 4 be quoted as **typical** of his work in view of the high marks obtained on three occasions.

The **mean** though is a measure which takes account of every item of data. In the example above the pupil has clearly been inconsistent in his work. If he had been consistent in his work what mark would he have had to obtain each time to achieve the same total mark for all seven pieces?

$$\text{Total mark} = 3+4+4+4+9+10+10 = 44$$

$$\text{Consistent mark} = \frac{44}{7} \approx 6.3$$

This is in fact the **arithmetic mean** of his marks and is what most people would describe as the **average mark.**

But what does the **mean** actually mean? The mean is the most commonly used of all the 'typical' values but often the least understood. The mean can be basically thought of as a balancing device. Imagine that weights were placed on a 10 cm bar in the places of the marks above. In order to balance the data the pivot would have to be placed at 6.3

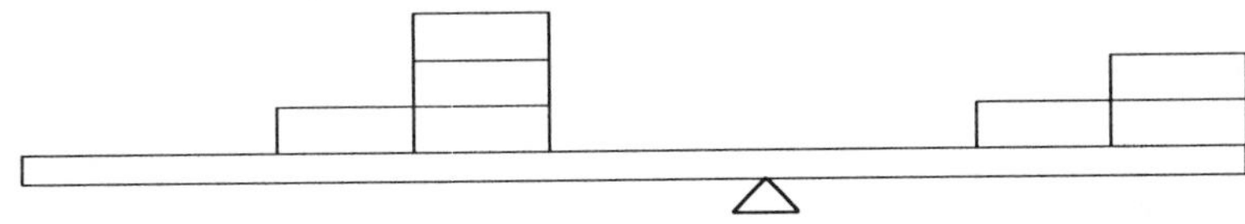

This is both the strength and weakness of the mean; whilst it uses all the data and takes into account end values it can easily be distorted by extreme values. For example, if in a small company the boss earns £30 000 per annum and his six workers £5000, then

$$\begin{aligned}\text{mean} &= \frac{1}{7}(30000+5000+5000+5000+5000+5000+5000)\\ &= £8571\end{aligned}$$

The workers might well argue however that this is **not** a typical wage at the company!

In general though, the mean of a set of data x_i i.e. $x_1, x_2, \ldots, x_n$ is given by

$$\bar{x} = \frac{\Sigma x_i}{n}$$

The summation is over i, but often for shorthand it is simply written as

$$\bar{x} = \frac{\Sigma x}{n}$$

Activity 6 What do you mean?

In the BBC 'Yes Minister' programme the Prime Minister instructs his Private Secretary to give the Press the average wage of a group of workers. The Private Secretary asks, 'Do you mean the wage of the average worker or the average of all the workers' wages?' The PM replies, 'But they are the same thing, aren't they?' Do you agree?

Exercise 3G

Employment in manufacturing

% of total civilian employment

	1960	1970	1971	1972	1973	1974	1975	1976	1977	1978	1979	1980	1981	1982	1983
Canada	23.7	22.3	21.8	21.8	22.0	21.7	20.2	20.3	19.6	19.6	19.9	19.7	19.3	18.1	17.5
US	27.1	26.4	24.7	24.3	24.8	24.2	22.7	22.8	22.7	22.7	22.7	2.1	21.7	20.4	19.8
Japan	21.5	27.0	27.0	27.0	27.4	27.2	25.8	25.5	25.1	24.5	24.3	24.7	24.8	24.5	24.5
France	27.5	27.8	28.0	28.1	28.3	28.4	27.9	27.4	27.1	26.6	26.1	25.8	25.1	24.7	24.3
W. Germany	37.0	39.4	37.4	36.8	36.7	36.4	35.6	35.1	35.1	34.8	34.5	34.3	33.6	33.1	32.5
Italy	23.0	27.8	27.8	27.8	28.0	28.3	28.2	28.0	27.5	27.1	26.7	26.7	26.1	25.7	24.7
Netherlands	30.6	26.4	26.1	25.6	25.4	25.6	25.0	23.8	23.2	23.0	22.3	21.5	20.9	20.5	20.3
Norway	25.3	26.7	25.3	23.8	23.5	23.6	24.1	23.2	22.4	21.3	20.5	20.3	20.2	19.7	18.2
UK	36.0	34.5	33.9	32.8	32.2	32.3	30.9	30.2	30.3	30.0	29.3	28.1	26.2	25.3	24.5

1. The information in the table above gives the percentage of workers employed in the manufacturing industry in the major industrial nations. Find the average percentage employed for 1960, 1975 and 1983. What does this tell you about the involvement of people in manufacturing industry in this period?

2. The results shown opposite are the final positions in the First Division Football in the 1990/91 season.

   (a) Total the goals scored both home and away and hence find the mean number of goals scored per match for each team.

   (b) Plot a scattergram of x, position in league, against y, average goals scored. How true is it that a high goal scoring average leads to a higher league position?

Division One

			Home					Away					
Pos		P	W	D	L	F	A	W	D	L	F	A	Pts
1	Arsenal	38	15	4	0	51	10	9	9	1	23	8	83
2	Liverpool	38	14	3	2	42	13	9	4	6	35	27	76
3	Crystal Pal	38	11	6	2	26	17	9	3	7	24	24	69
4	Leeds Utd	38	12	2	5	46	23	7	5	7	19	24	64
5	Man City	38	12	3	4	35	25	5	8	6	29	28	62
6	Man Utd	37	11	3	4	33	16	5	8	6	24	28	58
7	Wimbledon	38	8	6	5	28	22	6	8	5	25	24	56
8	Nottm For	38	11	4	4	42	21	3	8	8	23	29	54
9	Everton	38	9	5	5	26	15	4	7	8	24	31	51
10	Chelsea	38	10	6	3	33	25	3	4	12	25	44	49
11	Tottenham	37	8	9	2	35	22	3	6	9	15	27	48
12	QPR	38	8	5	6	27	22	4	5	10	17	31	46
13	Sheff Utd	38	9	3	7	23	23	4	4	11	13	32	46
14	Southptn	38	9	6	4	33	22	3	3	13	25	47	45
15	Norwich	38	9	3	7	27	32	4	3	12	14	32	45
16	Coventry	38	10	6	3	30	16	1	5	13	12	33	44
17	Aston Villa	38	7	9	3	29	25	2	5	12	17	33	41
18	Luton	38	7	5	7	22	18	3	2	14	20	43	37

(c) The table below gives, amongst other information, the mean 'Goals Scored' and 'Goals Conceded' for the successful years of Arsenal. What do these 'averages' tell you about the scores in matches of earlier years?

Seasons of success: How Arsenal's past and present League triumphs measure up

	Games							Average goals per match	
Season	**P**	**W**	**D**	**L**	**Pts**	**F**	**A**	**Scored**	**Conceded**
1990 - 91	38	24	13	1	83	74	18	1.95	0.48
1988 - 89	38	22	10	6	76	73	36	1.89	0.95
1970 - 71	42	29	7	6	85	71	29	1.69	0.69
1932 - 33	42	25	8	9	75	118	61	2.81	1.45

3. Find the mean playing time of the tracks of one of your albums. How does this compare with your median time? Which do you think is a better measure?

3.9 Using your calculator

Most modern calculators have a statistical function. This enables a running check to be kept on the total and number of results entered. Check your instruction booklet on how to do this. It is good practice when entering a set of values always to check the n memory to ensure you haven't missed a value out or put in too many. A common fault is to forget to clear a previous set of results.

When dealing with large amounts of data it is easy to make a mistake in adding up totals or entering. For example, the number of children in families for a class of children was recorded opposite:

No. of children (x)	**Frequency** (f)
1	8
2	11
3	6
4	4
5	1

The total could be found by repeated addition,

i .e $1+1+1+1+1+1+1+1+2+2 \ldots +4+4+4+4+5.$

However, it is far simpler to multiply the x values by the frequencies,

i.e. $(1\times8)+(2\times11)+(3\times6)+(4\times4)+(5\times1).$

So if n is the sum of the frequencies, in general

$$\bar{x} = \frac{\Sigma x_i f_i}{\Sigma f_i} \text{ when } n = \Sigma f_i$$

Most calculators can automatically enter frequencies - check your calculator instructions carefully.

With grouped frequency tables the same principle applies except that for the x value the mid-mark of the group is used (i.e. the value half way between the class limits). This is not entirely accurate as it assumes an **even spread** of data within the group. Usually differences above and below will cancel out but beware of quoting values with too high a degree of accuracy. The ages of people injured in road accidents in Cornwall in 1988 are shown opposite.

Age	Mid-mark	Frequency	$x \times f$
1 -10	5.5	199	1094.5
11-20	15.5	895	13872.5
21-30	25.5	625	15937.5
31-40	35.5	388	13774.0
41-50	45.5	261	11875.5
51-60	55.5	153	8491.5
61-70	65.5	141	9235.5
70+	75.5	140	10570.0
		2802	84851.0

This gives

$$\bar{x} = \frac{84851}{2802} \approx 30$$

Note that in the last open ended group a mid-mark of 75.5 was used to tie in with other groups. However, as this has a high frequency it could be a cause of error if there were, in fact, a significant number of over 80-year-olds involved in accidents.

Exercise 3H

1. The table opposite shows the wages earned by YTS trainees in 1984. Do you think that the mean of £28.10 is a fair figure to quote in these circumstances? What figure would you quote and why?
2. Find the mean number of shares issued by Cable & Wireless PLC as given in Exercise 3F, Q3. Why is there such a difference between the median and the mean? What information might be useful in obtaining a more accurate estimate of the mean?

Weekly income of trainees (March 1984)

Income	Per cent of trainees
£25.00	84
Over £25.00 up to £30.00	3
Over £30.00 up to £35.00	3
Over £35.00 up to £40.00	1
Over £40.00 up to £50.00	4
Over £50.00 up to £60.00	3
Over £60.00	2
	100

Mean £28.10

3.10 How spread out are the data?

Activity 7 Do differences in height even out as you get older?

Earlier you collected heights of people in your own age group. Collect at least 20 heights of people in an age group four or five years younger. Is there more difference in heights in the younger age group than in the older?

This section will examine ways of looking at this.

Example

Multiple discipline endurance events have gained in popularity over the last few years. The data on the next page gives the results of the first 50 competitors in a biathlon race consisting of a 15 mile bike ride followed by a 5 mile run. Some competitors argued that the race was biased towards cyclists as a good cyclist could make up more time in the cycling event which she or he would not lose on the shorter event. What you need to consider here is whether cycling times are more varied than running times.

Solution

The simplest way this could be done would be to look at the difference between the fastest and slowest times for each part. This is the **range**.

For cycling

$$\text{range} = 1\text{h } 9\text{s} - 44\text{min } 50\text{s} = 15\text{min } 19\text{s}$$

and for running

$$\text{range} = 48\text{min } 51\text{s} - 32\text{min } 23\text{s} = 16\text{min } 28\text{s}.$$

So, on the face of it, running times are more spread out than cycling times. However, in both sets of figures there are odd results at the end which differ by more than the difference of 2 minutes from the rest of the results and can on their own account for the difference in ranges. The range is therefore far too prone to effects of extremes and is of limited practical use.

Some statisticians use $\frac{n}{2}$ for the median, $\frac{n}{4}$, $\frac{3n}{4}$ for the quartiles when using grouped data – this is acceptable, and would not be penalised in the AEB Statistics Examination.

To overcome this, the **inter-quartile range** (IQR) attempts to miss out these extremes. The **quartiles** are found in the same way as the median but at the $\frac{(n+1)}{4}$ th and $\frac{3(n+1)}{4}$ th item of data.

Taking just the fastest seven items of cycling data, look for the quartiles at the 2nd and 6th item:

44:50	45:25	47:15	47:16	48:07	48:07	48:18
	↑		↑		↑	
	lower quartile (LQ)		median		upper quartile (UQ)	

The inter-quartile range $= 48.07 - 45.25 = 2\text{min } 42\text{s}.$

This tells you the range within which the middle 50% of data lies. In some cases, where the data are roughly symmetrical, the **semi inter-quartile range** is used. This gives the range either side of the median which contains the middle 50% of data.

Mildenhall C.C.
Biathalon 30.8.87
Results

Finishing order Position	No	Name	Club	Cycle Time	Run Time	Total Time
1	157	Roy E. Fuller	Ely & Dist C.C.	48.18	33.55	1.22.13
2	106	Clive Catchpole	Fitness Habit (Ipswich)	45.25	36.59	1.22.24
3	108	Robert Quarton	Fitness Habit (Ipswich)	48.50	33.45	1.22.35
4	26	Michael Bennett	Fitness Habit (Ipswich)	47.15	35.47	1.23.02
5	110	David Minns	West Suffolk A.C. Mildenhall C.C/Dairytime	51.00	32.32	1.23.32
6	30	Christopher Neale	Surrey Road C.C.	48.07	36.33	1.24.40
7	46	Roger Jackerman	Met Police A.A.	50.15	35.14	1.25.29
8	60	David Chamborlain	Scalding C.C. Holbeach A.C.	48.07	37.39	1.25.46
9	66	Nigel Morrison	Halstead Roadrunners	48.50	37.15	1.26.05
10	80	Michael Meyer		49.50	37.04	1.26.54
11	143	Paul Chapman	Bishop Stortford C.C.	50.00	37.10	1.27.10
12	120	Chris Carter	North Bucks R.C.	47.16	39.57	1.27.13
13	123	Ian Coles	Colchester Rovers	49.55	37.43	1.27.38
14	102	Stephen Nobbs	North Norfolk Beach Runners	53.12	34.42	1.27.54
15	171	David Smith	Ipswich Jaffa	55.46	32.23	1.28.09
16	129	Don Hutchinson	Sir M. McDonald & Partners Running Club	52.03	36.08	1.28.11
17	50	Bill Morgan	Diss & Dist Wheelers	49.15	37.46	1.29.01
18	169	C. Willmets	Cambridge Triathlon	50.45	38.32	1.29.51
19	155	John Wright	Duke St. Runners	55.25	34.11	1.29.36
20	58	R. F. Williams	North Norfolk Beach Runners	52.50	37.01	1.29.51
21	187	Jon Trevor	East London Triathletes Unity C.C.	51.30	38.22	1.29.52
22	18	Julian Tomkinson		55.12	34.55	1.30.07
23	181	G. Carpenter		58.15	32.38	1.30.53
24	56	Duncan Butcher	St. Edmund Pacers	55.42	35.18	1.31.00
25	147	H. D. Ward	Colchester Rovers	49.45	41.39	1.31.24
26 =	40	Jeffrey P. Hathaway	North Bucks R.C.	44.50	46.51	1.31.41
26 =	12	Steven Elvin		55.15	36.26	1.31.41
28	165	Geoffrey Davidson	Wymondham Joggers	53.00	38.43	1.31.43
29	175	Mike Parkin	Deeping C.C.	50.35	41.50	1.32.35
30	149	Pete Cotton	Mildenhall C.C./Dairytime	54.25	38.21	1.32.46
31	84	Barry Parker	Thetford A.C. Wymondham Joggers	53.48	39.17	1.33.05
32	90	Keith Tyler	Wisbech Wheelers Cambs Speed Skaters	48.45	44.54	1.33.39
33	36	Derek Ward	Duke St. Runners	54.10	39.41	1.33.51
34	38	Gordon Bidwell	West Norfolk A.C.	55.17	38.36	1.33.53
35	139	John M. Chequer	Granta Harriers	54.35	39.55	1.34.30
36	59	Jeremy Hunt	ABC Centerville	53.20	41.5	1.34.35
37	133	W. E. Clough	Cambridge Town & County C.C.	52.32	42.22	1.34.54
38	163	Bruce Short	West Norfolk Rugby Union	51.10	44.02	1.35.12
39	185	Kate Byrne	East London Triathletes Unity C.C.	54.05	41.17	1.35.22
40	29	Justin Newton	Mildenhall C.C./Dairytime	56.20	40.54	1.37.14
41	127	S. Kennett		58.40	38.45	1.37.25
42	14	David J. Cassell	Bungay Black Dog	57.59	40.11	1.38.10
43	78	Roger Temple		54.27	44.26	1.38.53
44	141	Lulu Goodwin		53.37	45.37	1.39.14
45	48	Patrick Ash	North Norfolk Beach Runners North Norfolk Wheelers	55.27	44.06	1.39.33
46	62	Philip Mitchell		55.54	43.44	1.39.38
47	76	Parry Pierson Cross	Havering C. T. C.	50.48	48.51	1.39.39
48	118	Geoff Holland	Wymondham Joggers	57.12	42.44	1.39.56
49	197	Terry Scott		1.00.09	40.01	1.40.10
50	137	Nigel Chapman	Bishop Stortford C.C.	57.45	42.33	1.40.18

With grouped data you can use either the interpolation method or a cumulative frequency curve to find the quartiles and hence the IQR. For cycling, the graphed data are summarised opposite.

Cycling Times	Frequencies	Cumulative Frequency
44:00-45:59	2	2
46:00-47:59	2	4
48:00-49:59	10	14
50:00-51:59	8	22
52:00-53:59	8	30
54:00-55:59	13	43
56:00-57:59	4	47
58:00 +	3	50

The cumulative frequency curve is shown below. Note that you plot (46, 2), (48, 4), etc. but that the last point cannot from this grouped data be plotted.

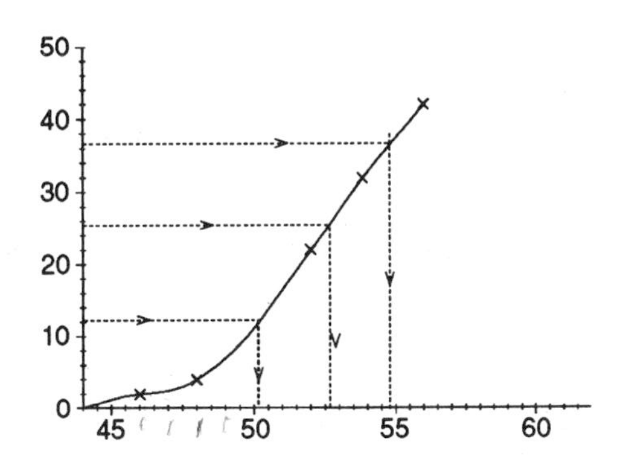

The median is given by the

$$\frac{(50+1)}{2} = 25.5 \text{ th}$$

item of data. So drawing across to the cumulative frequency curve and then downwards gives an estimate of the median as 52.6.

Similarly estimates for the quartiles are given by the

$$\frac{(50+1)}{4} = 12.75\text{th item}$$

and the $$\frac{3(50+1)}{4} = 38.25\text{th item.}$$

This gives estimates

$$\text{LQ} = 50.1 \text{ min}, \quad \text{UQ} = 54.8 \text{ min}$$

with an inter-quartile range of $54.8 - 50.1 = 4.7$ min.

Using interpolation, the lower quartile is at the 12.75th item, and an estimate for this, since there are 4 items up to 48:00 and 10 items in the next group which has class width 2, is given by

$$\text{LQ} = 48.0 + \left[\frac{(12.75-4)}{10} \times 2\right]$$

$$= 49.8 \text{ min}.$$

Similarly the upper quartile is the 38.25 th item, and an estimate is

$$\text{UQ} = 54.00 + \left[\frac{(38.25-30)}{13}\times 2\right]$$

$$= 55.3 \text{ min.}$$

Hence the inter-quartile range is given by

$$\text{IQR} = 55.3 - 49.8 = 5.5 \text{ min}.$$

If a stem and leaf diagram has been used, the median and quartiles can be taken from the data directly. To assist in this, the cumulative frequencies are calculated working from both ends to the middle. The stem and leaf diagram for the **rounded decimal times** is shown opposite. The stem is in minutes, and the leaf is rounded to one d.p. of a minute.

(1)	44	8	
(2)	45	4	
(2)	46		
(4)	47	33	
(10)	48	113888	
(14)	49	3(88)9	Lower quartile
(19)	50	03688	
(22)	51	025	
(25)	52	15(8)	
(25)	53	(0)368	Median
(21)	54	12456	
(16)	55	233(45)7899	Upper quartile
(7)	56	3	
(6)	57	28	
(4)	58	137	
(1)	59		
(1)	60	2	

A new form of diagram, using the median and quartiles, is becoming increasingly popular . The **box and whisker plot** shows the data on a scale and is very useful for comparing the 'distribution' of several sets of data drawn on the same scale.

The box is formed by using the two quartiles, and the median is illustrated by a cross. The whiskers are found by using minimum and maximum values, as illustrated below.

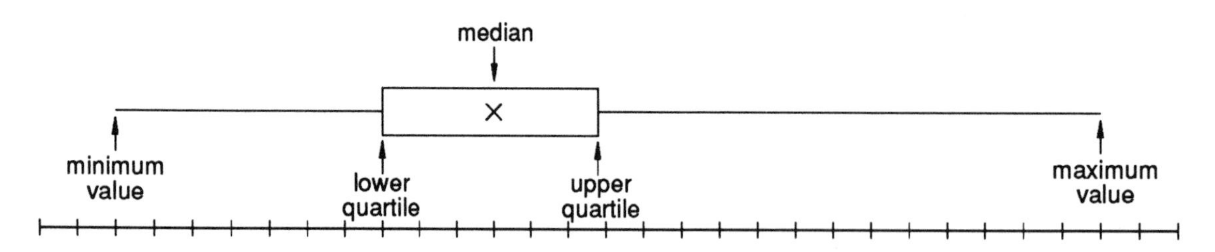

Example

Use a box and whisker plot to illustrate the following two sets of data relating to exam results of 11 candidates in Mathematics and English.

Pupil	A	B	C	D	E	F	G	H	I	J	K
Maths	62	91	43	31	57	63	80	37	43	5	78
English	65	57	55	37	62	70	73	49	65	41	64

Solution

Rearrange each set of data into increasing order.

Maths 5 31 37 43 43 57 62 63 78 80 91

↑ LQ ↓ ↑ median ↓ ↑ UQ ↓

English 37 41 49 55 57 62 64 65 65 70 73

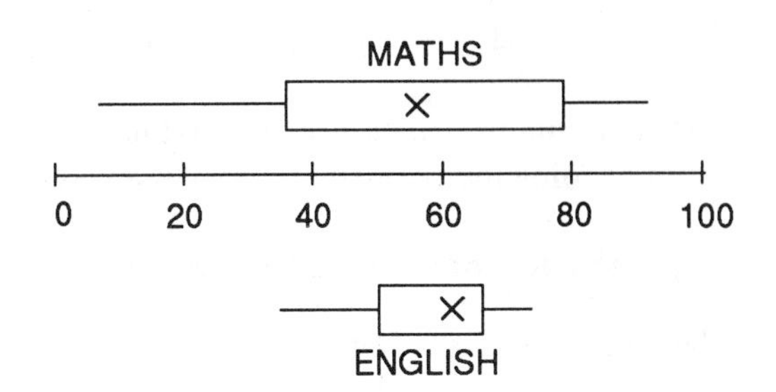

This diagram helps you to see quickly the main characteristics of the data distribution for each set. It does not, however, enable comparisons to be made of the relative performances of candidates.

Exercise 3I

1. Using any method find the IQR of the running times shown in the table of biathlon results at the start of this section. Are the competitors justified in their complaint?
2. Find the median and IQR for the heights of both age groups measured in earlier activities. Are heights more varied at a particular age?
3. When laying pipes, engineers test the soil for 'resistivity'. If the reading is low then there is an increasing risk of pipes corroding. In a survey of 159 samples the following results were found:

Resistivity (ohms/cm)	Frequency
400 - 900	5
901 - 1500	9
1501 - 3500	40
3501 - 8000	45
8001 - 20000	60

Find the median and inter-quartile range of this data.

3.11 Standard deviation

Like the median, the quartiles fail to make use of all the data. This can of course be an advantage when there are extreme items of data. There is a need then for a measure which makes use of **all** data. There is also a need for a measure of **spread** which relates to a central value. For example, two classes who sat the same exam might have the same mean mark but the marks may vary in a different pattern around this. It seems sensible if you are using all the data that the measure of spread ought to be related to the mean.

One method sometimes used is the **mean deviation from the mean.**

For example, take the following data:

6, 8, 8, 9, 14, 15,

the mean of which is 10.

The differences, or deviations, of these from the mean are given by

$$-4,\ -2,\ -2,\ -1,\ +4,\ +5.$$

To find a summary measure you first need to combine these, but by simply adding them together you will always get zero.

Why is the sum of the deviations always zero?

The mean deviation simply ignores the sign, using what is known in mathematics as the **modulus**, e.g. $|-3| = 3$ and $|3| = 3$. In order that the measure is not linked to the size of sample, you then average the deviations out:

$$\text{mean deviation from the mean} = \frac{1}{n}\Sigma\left|x_i - \bar{x}\right|$$

In the example, this has value $\frac{1}{6}(4+2+2+1+4+5) = 3$.

However, just ignoring signs is not a very sound technique and the mean deviation is not often used in practice.

Activity 8 Pulse rates

The pulse rates of a group of 10 people were:

72, 80, 67, 68, 80, 68, 80, 56, 76, 68.

The mean of this data is about 70. Now calculate the deviations of all the values from this 'assumed' mean. Instead of just ignoring the signs however, square the deviations and add these together,

i.e $\quad 2^2 + 10^2 + 3^2 + 2^2 + 10^2 + 2^2 + 10^2 + 14^2 + 6^2 + 2^2 = 557$

Note how the sign now becomes irrelevant.

Repeat this with other assumed means around the same value and put the results in a table (it will save time to work in a group):

Assumed mean	67	68	69	69.5	70	70.5	71	72	73
Σd^2					557				

Now plot a graph of these results.

What you should find in this activity is that the results form a quadratic graph. The value of assumed mean at the bottom of the graph is the value for which the sum of the squared deviations is the least. Find the arithmetic mean of your data and you may not be surprised to find that this is the same value. This idea is an important one in statistics and is called the **'least squares method'**.

Squaring the deviations then is an alternative to using the modulus and the result can be averaged out over the number of items of data. This is known as the **variance**. However, the value can often be disproportionately large and it is more common to square root the variance to give the **standard deviation**. So

$$\text{variance} \quad s^2 = \frac{1}{n}\Sigma(x_i - \bar{x})^2$$

$$\text{standard deviation} \quad s = \sqrt{\frac{1}{n}\Sigma(x_i - \bar{x})^2}$$

Example

Find the standard deviation of the pulse rates in Activity 8.

Solution

$\bar{x} = 71.6$, so you have the following table:

	72	80	67	68	80	68	80	56	76	69
$\lvert x-\bar{x}\rvert$	0.4	8.4	4.6	3.6	8.4	3.6	8.4	15.6	4.4	2.6
$(x-\bar{x})^2$	0.16	70.6	21.2	13.0	70.6	13.0	70.6	243.4	19.4	6.8

giving $\Sigma(x - \bar{x})^2 = 528.76$.

Hence variance, $s^2 = \dfrac{528.76}{10} \approx 52.9$,

and standard deviation, $s \approx 7.27$.

It is very tedious to calculate by this method - even using a calculator you would have problems, as the calculator would have to memorise all the data until the mean could be calculated. An alternative formula often used is

$$s^2 = \left(\frac{1}{n}\Sigma x^2\right) - \bar{x}^2$$

You can derive this result by noting that

$$\begin{aligned} s^2 &= \frac{1}{n}\Sigma(x_i - \bar{x})^2 \\ &= \frac{1}{n}\Sigma(x_i^2 - 2x_i\bar{x} + \bar{x}^2) \\ &= \frac{1}{n}\Sigma x_i^2 - \frac{2\bar{x}}{n}\Sigma x_i + \frac{\bar{x}^2}{n}\Sigma 1. \end{aligned}$$

But $\quad \frac{1}{n}\Sigma x_i = \bar{x}$ and $\Sigma 1 = n$,

giving $\quad s^2 = \frac{1}{n}\Sigma x_i^2 - 2\bar{x}^2 + \bar{x}^2$

or $\quad s^2 = \frac{1}{n}\Sigma x_i^2 - \bar{x}^2.$

Calculators use this method and keep a running total of

(a) n the quantity of data entered,

(b) Σx the running total,

(c) Σx^2 the sum of the values squared.

x	Σx	Σx^2
72	72	5184
80	152	11584
67	219	16073
..	..	..
..	..	..
..	..	..
69	716	51794

This is illustrated opposite, and

$$\bar{x} = \frac{716}{10} = 71.6$$

$$s = \sqrt{\frac{51794}{10} - 71.6^2} = 7.27.$$

Find out how to use your calculator to calculate the standard deviation (SD). Most will give you all the values in the above formula too.

What does the standard deviation stand for?

Whereas you were able to say that the IQR was the range within which the middle 50% of a data set lies there is no absolute meaning that can be given to the SD. On its own then it can be difficult to judge the significance of a particular SD.

It is of more use to compare two sets of data.

Example

Compare the means and standard deviation of the two sets of data

(a) 3, 4, 5, 6, 7

(b) 1, 3, 5, 7, 9

Solution

(a) $$\bar{x} = \frac{3+4+5+6+7}{5} = 5,$$

and $$s^2 = \frac{1}{5}(9+16+25+36+49) - 25$$

$$= 27 - 25 = 2,$$

giving $s \approx 1.414$.

(b) As in (a), $\bar{x} = 5$,

but $$s^2 = \frac{1}{5}(1+9+25+49+81) - 25$$

$$= 33 - 25 = 8,$$

giving $s \approx 2.828$.

Thus the two sets of data have equal means but since the spread of the data is very different in each set, they have different SDs. In fact, the second SD is double the first.

Activity 9

Construct a number of data sets similar to those in the example, which all have the same means. Estimate what you think the standard deviation will be. Now calculate the values and see if they agree with your intuitive estimate.

Activity 10

Find the standard deviation of the album track length data used earlier. Do some albums have more varied track lengths than others?

With grouped frequency tables the SD can be calculated as follows. Find Σx and Σx^2 by multiplying the frequency by the mid-marks and the mid-marks squared respectively.

e.g.

Height	Frequency	Σx	Σx^2
140-149	5	5×144.5	$5 \times (144.5)^2$

As with means, most modern calculators can perform these operations in statistical mode.

Example

The lengths of 32 fish caught in a competition were measured correct to the nearest mm. Find the mean length and the standard deviation.

Length	20-22	23-25	26-28	29-31	32-34
Frequency	3	6	12	9	2

Solution

Group	Mid-point (x)	Frequency (f)	fx	$f(x^2)$
20-22	21	3	63	1323
23-25	24	6	144	3456
26-28	27	12	324	8748
29-31	30	9	270	8100
32-34	33	2	66	2178
		$\Sigma f = 32$	$\Sigma fx = 867$	$\Sigma fx^2 = 23805$

So $$\bar{x} = \frac{\Sigma x_i}{n} = \frac{\Sigma f x}{\Sigma f} = \frac{867}{32} \approx 27.1$$

and $$s^2 = \frac{\Sigma x_i^2}{n} - \bar{x}^2 = \frac{\Sigma f x^2}{\Sigma f} - \bar{x}^2$$

$$= \frac{23805}{32} - \left(\frac{867}{32}\right)^2 \approx 9.835$$

$$\Rightarrow \quad s \approx 3.14$$

Note that, for grouped data, the general formulae for mean and standard deviation became

$$\bar{x} = \frac{\Sigma f x}{\Sigma f}, \quad s^2 = \frac{\Sigma f x^2}{\Sigma f} - \bar{x}^2.$$

Exercise 3J

1. From the frequency tables drawn up earlier for the biathlon race find the standard deviations of the running and cycling times. Are cycling times more varied?
2. The data opposite give the age of mothers of children born over the last 50 years. Find the mean and SD of the ages for 1941, 1961 and 1989. What does this tell you about the change in the age at which women are tending to have children?

Live births: by age of mother

Great Britain — Percentages

Age of mother	1941	1951	1961	1971	1981	1989
15-19	4.3	4.3	7.2	10.6	9.0	8.2
20-24	25.4	27.6	30.8	36.5	30.9	26.9
25-29	31.0	32.2	30.7	31.4	34.0	35.4
30-34	22.1	20.7	18.8	14.1	19.7	21.1
35-39	12.7	11.5	9.6	5.8	5.3	7.0
40-44	4.2	3.4	2.7	1.5	1.0	1.3
45-49	0.3	0.2	0.2	0.1	0.1	0.1

(Source: Population Censuses and Surveys Scotland)

3. The data below give the usual working hours of men and women, both employed and self-employed. Find the mean and standard deviation of the four groups and use this information to comment on the differences between men and women and employed/self-employed people.

Basic usual hours worked: by sex and type of employment, 1989

Great Britain				**Percentages**
	Males		**Females**	
	Employees	**Self employed**	**Employees**	**Self employed**
Hours per week				
Less than 5	0.4	1.0	2.2	6.0
5 but less than 10	1.1	0.9	6.5	7.3
10 but less than 15	1.0	1.1	7.8	9.2
15 but less than 20	0.7	0.9	9.4	7.4
20 but less than 25	0.9	1.6	10.9	8.5
25 but less than 30	1.0	1.3	5.9	5.4
30 but less than 35	2.6	3.2	6.9	7.7
35 but less than 40	50.7	8.6	38.7	9.1
40 but less than 45	28.6	26.0	9.1	13.1
45 but less than 50	5.2	12.5	1.0	6.3
50 but less than 55	3.0	12.7	0.6	4.4
55 but less than 60	1.3	4.6	0.2	2.4
60 and over	3.2	25.2	0.6	12.8

(Source: Labour Force Survey Employment Department)

3.12 Miscellaneous Exercises

1. The data below show the length of marriages ending in divorce for the period 1961-1989. Using the data for 1961, 1971, 1981 and 1989:
   (a) draw any diagrams which you think useful to illustrate the pattern of marriage length;
   (b) calculate any measures which you think appropriate;
   (c) write a short report on the pattern of marriage breakdowns over this period.

Percentages and thousands

Year of divorce	**1961**	**1971**	**1976**	**1981**	**1983**	**1984**	**1985**	**1986**	**1987**	**1988**	**1989**
Duration of marriage (percentages)											
0-2 years	1.2	1.2	1.5	1.5	1.3	1.2	8.9	9.2	9.3	9.5	9.8
3-4 years	10.1	12.2	16.5	19.0	19.5	19.6	18.8	15.3	13.7	13.4	13.4
5-9 years	30.6	30.5	30.2	29.1	28.7	28.3	36.2	27.5	28.6	28.0	28.0
10-14 years	22.9	19.4	18.7	19.6	19.2	18.9	17.1	17.5	17.5	17.5	17.6
15-19 years	13.9	12.6	12.8	12.8	12.9	13.2	12.2	12.8	13.0	13.2	13.0
20-24 years		9.5	8.8	8.6	8.6	8.7	7.9	8.4	8.7	9.1	9.0
25-29 years	21.2	5.8	5.6	4.9	5.2	5.3	4.7	4.8	4.9	4.9	4.9
30 years and over		8.9	5.9	4.5	4.7	4.6	4.2	4.3	4.3	4.3	4.3
All durations (= 100%) (thousands)	27.0	79.2	134.5	155.6	160.7	156.4	173.7	166.7	163.1	164.1	162.5

2. As a result of examining a sample of 700 invoices, a sales manager drew up the grouped frequency table of sales shown opposite.
   (a) Calculate the mean and the standard deviation of the sample.
   (b) Explain why the mean and the standard deviation might not be the best summary statistics to use with these data.
   (c) Calculate estimates of alternative summary statistics which might be used by the sales manager. Use these estimates to justify your comment in (b).

Amount on invoice (£)	**Number of invoices**
0-9	44
10-19	194
20-49	157
50-99	131
100-149	69
150-199	40
200-499	58
500-749	7

3. Using the number of incomes in each category, calculate the mean income in 1983/4 and 1984/5.

   Do you think these are the best measures to use here? Give your reasons and suggest alternative measures.

1983/84 Annual Survey

Lower limit of range of income	Thousands Number of incomes
All incomes	**22 015**
Income before tax £	
1 500	509
2 000	1 230
2 500	1 070
3 000	1 200
3 500	1 220
4 000	1 240
4 500	1 130
5 000	1 140
5 500	1 100
6 000	1 890
7 000	1 710
8 000	2 810
10 000	2 040
12 000	1 740
15 000	1 120
20 000	645
30 000	169
50 000	44
1000 000 and over	8

1984/85 Annual Survey

Lower limit of range of income	Thousands Number of incomes
All incomes	**22 164**
Income before tax £	
2 000	1 340
2 500	1 000
3 000	1 060
3 500	1 090
4 000	1 210
4 500	1 090
5 000	1 060
5 500	1 985
6 000	1 190
7 000	1 690
8 000	2 930
10 000	2 090
12 000	1 990
15 000	1 340
20 000	780
30 000	246
50 000	62
1000 000 and over	11

4. The table opposite shows the lifetimes of a random sample of 200 mass produced circular abrasive discs.

   (a) Without drawing the cumulative frequency curve, calculate estimates of the median and quartiles of these lifetimes.

   (b) One method of estimating the skewness of a distribution is to evaluate

   $$\frac{3\,(\text{mean} - \text{median})}{\text{standard deviation}}.$$

   Carry out the evaluation for the above data and comment on your result.

   Use the quartiles to verify your findings.

Lifetime (to nearest hour)	Number of discs
690-709	3
710-719	7
720-729	15
730-739	38
740-744	41
745-749	35
750-754	21
755-759	16
760-769	14
770-789	10

5. The following information is taken from a government survey on smoking by schoolchildren.

Cigarette consumption (per week)	**England and Wales 1982**	**1984**	**1986**
Boys	%	%	%
None	12	13	12
1-5	24	24	25
6-40	33	31	30
41-70	16	16	18
71 and over	16	14	15
Mean	33	31	33
Median	15	16	20
Base (= 100%)	*272*	*419*	*210*
Girls			
None	13	10	10
1-5	29	26	21
6-40	32	34	38
41-70	14	15	16
71 and over	11	14	15
Mean	26	30	32
Median	11	14	17
Base (= 100%)	*289*	*373*	*266*

(a) Both the mean and median have been calculated for each category. Why do these differ so much? Which would you prefer as a suitable measure in this survey?

(b) Write a short report using suitable illustrations on the pattern of teenage smoking over the years 1982-1986.

6. The data below form part of a survey on the TV watching habits of schoolchildren.

(a) Find the mean and SD for boys and girls in each age group and comment on any differences.

(b) By combining the boys' and girls' standard deviations and means, assuming an equal number of each took part in the survey, find overall figures for each age group.

7. In order to monitor whether large firms are taking over from smaller ones the government carries out a survey on company size at regular intervals. The results of such a survey are shown below.

(a) Draw a relative frequency histogram of the data.

(b) Calculate the mean and standard deviation of the size of companies.

(c) Find the median and quartiles of the data and use these to draw a box and whisker plot.

(d) Comment on the suitability of the measures in (b) and (c) and any inaccuracies in the calculation techniques.

Size bands according to numbers of employees	**Census units numbers**	**%**
1-10	847 537	73.6
11-24	169 800	14.7
25-49	70 671	6.1
50-99	32 888	2.9
100-199	17 236	1.5
200-499	9 352	0.8
500-999	2 605	0.2
1000+	1 476	0.1
Total	1 151 565	100.0

(Source: Department of Employment, Statistics Division, 1988)

8. 38 children solved a simple problem and the time taken by each was noted.

Time (seconds)	5-	10-	20-	25-	40-	45-
Frequency	2	12	7	15	2	0

Draw a histogram to illustrate this information.

	1st year(11+) Boys	**Girls**	**3rd year(13+) Boys**	**Girls**	**5th year(15+) Boys**	**Girls**
None	5.3	6.6	4.9	6.0	6.9	8.1
Less than 1hr	13.6	16.9	12.7	16.5	14.4	19.2
1-2hr	20.4	23.4	18.8	21.7	20.8	22.7
2-3hr	19.4	18.4	21.7	18.4	21.0	20.0
3-4hr	14.6	15.0	18.1	16.7	16.1	14.9
4-5hr	11.3	9.3	9.7	9.8	10.3	7.5
5hrs or longer	15.4	10.4	14.1	10.8	10.3	7.6

9. The number of passengers on a certain regular weekday train service on each of 50 occasions was:

165	141	163	153	130	158	119	187	185	209
177	147	166	154	159	178	187	139	180	143
160	185	153	168	189	173	127	179	163	182
171	146	174	149	126	156	155	174	154	150
210	162	138	117	198	164	125	142	182	218

Choose suitable class intervals and reduce these data to a grouped frequency table.

Plot the corresponding frequency polygon on squared paper using suitable scales. (AEB)

10. The percentage marks of 100 candidates in a test are given in the following tables:

No. of marks	0-19	20-29	30-39	40-49
No. of candidates	5	6	13	22

No. of marks	50-59	60-69	70-79	80-89
No. of candidates	24	16	8	6

Draw a cumulative frequency curve.

Hence estimate

(i) the median mark,

(ii) the lower quartile,

(iii) the upper quartile. (AEB)

11. The number of passengers on a certain regular weekday bus was counted on each of 60 occasions. For each journey, the number of passengers in excess of 20 was recorded, with the following results.

15	6	13	8	9	12	8	11	5	12
7	11	7	11	10	10	7	9	14	10
6	7	9	12	13	9	8	8	12	14
9	10	11	13	8	8	8	11	8	13
12	14	13	7	8	6	11	10	15	10
8	13	7	12	9	10	9	8	11	9

(a) Construct a frequency table for these data.

(b) Illustrate graphically the distribution of the number of passengers per bus.

(c) For this distribution state the value of

(i) the mode,

(ii) the range. (AEB)

12. The breaking strengths of 200 cables, manufactured by a specific company, are shown in the table below.

Plot the cumulative frequency curve on squared paper.

Hence estimate

(a) the median breaking strength,

(b) the semi inter-quartile range,

(c) the percentage of cables with a breaking strength greater than 2300 kg.

Breaking strength	Frequency (in 100s of kg)
0-	4
5-	48
10-	60
15-	48
20-	24
20-30	16

13. The gross registered tonnages of 500 ships entering a small port are given in the following table.

Gross registered tonnage (tonnes)	No. of ships
0-	25
400-	31
800-	44
1200-	57
1600-	74
2000-	158
3000-	55
4000-	26
5000-	18
6000- 8000	12

Plot the percentage cumulative frequency curve on squared paper.

Hence estimate

(a) the median tonnage,

(b) the semi inter-quartile range,

(c) the percentage of ships with a gross registered tonnage exceeding 2500 tonnes.

(AEB)

14. The following table refers to all marriages that ended in divorce in Scotland during 1977. It shows the age of the wife at marriage.

Age of wife (years)	16-20	21-24	25-29	30/over
Frequency	4966	2364	706	524

(Source: Annual Abstract of Statistics, 1990)

(a) Draw a cumulative frequency curve for these data.

(b) Estimate the median and the inter-quartile range.

The corresponding data for 1990 revealed a median of 21.2 years and an inter-quartile range of 6.2 years.

(c) Compare these values with those you obtained for 1977. Give a reason for using the median and inter-quartile range, rather than the mean and standard deviation for making this comparison.

The box-and-whisker plots below also refer to Scotland and show the age of the wife at marriage. One is for all marriages in 1990 and the other is for all marriages that ended in divorce in 1990. (The small number of marriages in which the wife was aged over 50 have been ignored.)

Age of wife at marriage, Scotland

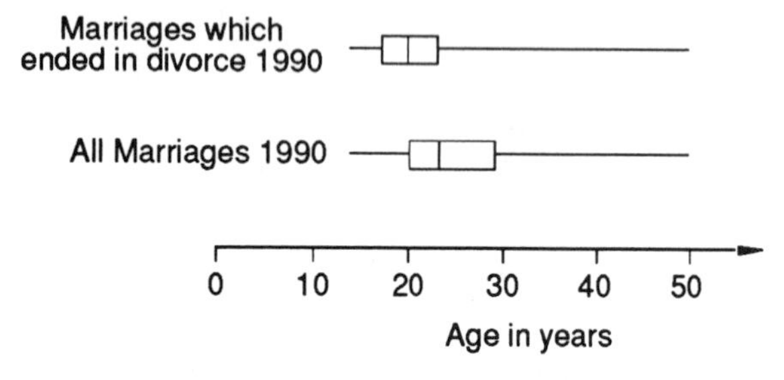

(d) Compare and comment on the two distributions. (AEB)

15. Give one advantage and one disadvantage of grouping data into a frequency table.

The table shows the trunk diameters, in centimetres, of a random sample of 200 larch trees.

Diameter (cm)	15-	20-	25-	30-	35-	40-50
Frequency	22	42	70	38	16	12

Plot the cumulative frequency curve of these data.

By use of this curve, or otherwise, estimate the median and the inter-quartile range of the trunk diameters of larch trees.

A random sample of 200 spruce trees yield the following information concerning their trunk diameters, in centimetres.

Min	Lower quartile	Median	Upper quartile	Max
13	27	32	35	42

Use this data summary to draw a second cumulative frequency curve on your graph.

Comment on any similarities or differences between the trunk diameters of larch and spruce trees. (AEB)

16. Over a period of four years a bank keeps a weekly record of the number of cheques with errors that are presented for payment. The results for the 200 accounting weeks are as follows.

Number of cheques with errors (x)	Number of weeks (f)
0	5
1	22
2	46
3	38
4	31
5	23
6	16
7	11
8	6
9	2

$$\left(\Sigma f x = 706 \quad \Sigma f x^2 = 3280\right)$$

Construct a suitable pictorial representation of these data.

State the modal value and calculate the median, mean and standard deviation of the number of cheques with errors in a week.

Some textbooks measure the **skewness** (or asymmetry) of a distribution by

$$\frac{3(\text{mean} - \text{median})}{\text{standard deviation}}$$

and others measure it by

$$\frac{(\text{mean} - \text{mode})}{\text{standard deviation}}.$$

Calculate and compare the values of these two measures of skewness for the above data.

State how this skewness is reflected in the shape of your graph.

(AEB)

17. Each member in a group of 100 children was asked to do a simple jigsaw puzzle. The times, to the nearest five seconds, for the children to complete the jigsaw are as follows:

Time (seconds)	60-85	90-105	110-125	130-145	150-165	170-185	190-215
No. of children	7	13	25	28	20	5	2

(a) Illustrate the data with a cumulative frequency curve.

(b) Estimate the median and the inter-quartile range.

(c) Each member of a similar group of children completed a jigsaw in a median time of 158 seconds with an inter-quartile range of 204 seconds. Comment briefly on the relative difficulty of the two jigsaws.

In addition to the 100 children who completed the first jigsaw, a further 16 children attempted the jigsaw but gave up, having failed to complete it after 220 seconds.

(d) Estimate the median completion time of the whole group of 116 children.

Comment on the use of the median instead of the arithmetic mean in these circumstances.

(AEB)

4 DISCRETE PROBABILITY DISTRIBUTIONS

Objectives

After studying this chapter you should

- understand what is meant by a discrete probability distribution;
- be able to find the mean and variance of distribution;
- be able to use the uniform distribution.

4.0 Introduction

The definition

'X = the total when two standard dice are rolled'

is an example of a random variable, X, which may assume any of the values in the range 2, 3, 4, ..., 12. The outcome cannot be predicted with certainty though probabilities can be assigned to each possible result.

A **random variable** is a quantity that may take any of a given range of values that cannot be predicted exactly but can be described in terms of their probability. As was seen in Chapter 2, data is classified either as **discrete** if the values are taken from a fixed number of numerical values (generally assessed by counting), or **continuous** if the values can fall anywhere over a range and the scale is only restricted by the accuracy of measuring. Some examples of data which can be described by a random variable are shown below.

Discrete	Continuous
number of red smarties in a packet	weight of babies at birth
number of traffic accidents in Leeds in one day	lengths of pine cones in a wood
number of throws required to score 6 with a single die	time needed to drive from Lincoln to Dover

Discuss whether the times taken to run 100 m in the Olympics will be values of a discrete rather than a continuous random variable.

4.1 Expectation

Activity 1

Play a game in which a counter is moved forward one, two or four places according to whether the scores on the two dice rolled differ by three or more, by one or two, or are equal. Here is a random variable, M, the number of places moved, which can take the values 1, 2 or 4. Play the game at least 20 times and evaluate from the games the average (mean) number of moves per game.

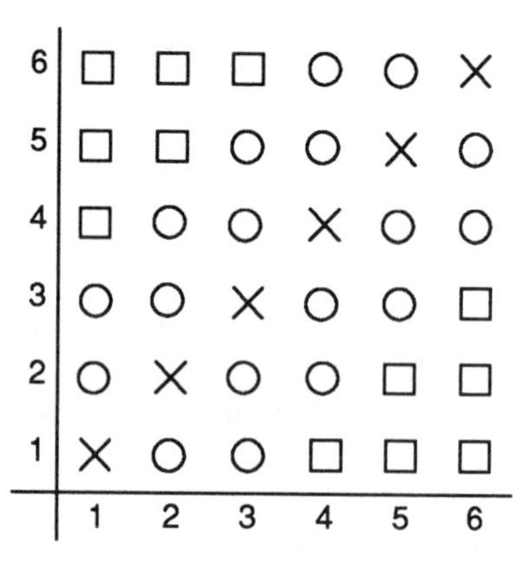

The probabilities of each of these values occurring can be calculated from the diagram opposite. Check for yourself that the probabilities are in fact,

$$P(M=1) = \frac{12}{36} = \frac{1}{3}$$

$$P(M=2) = \frac{18}{36} = \frac{1}{2}$$

$$P(M=4) = \frac{6}{36} = \frac{1}{6}.$$

In the long run you would expect to move one square $\frac{1}{3}$ of the times, two squares on $\frac{1}{2}$ of the goes and four squares on $\frac{1}{6}$. So if you play the game 36 times you will expect to average a total of

$$1\times12 + 2\times18 + 4\times6 = 12 + 36 + 24$$

$$= 72 \text{ moves.}$$

If you divide by 36 to get the mean number of moves per game the equation becomes

$$\frac{1\times12 + 2\times18 + 4\times6}{36} = \frac{72}{36}$$

which you can write as

$$1\times\frac{12}{36} + 2\times\frac{18}{36} + 4\times\frac{6}{36} = 2.$$

So mean $= 1\times P(M=1) + 2\times P(M=2) + 4\times P(M=4)$

$= 2.$

Using the summation symbol , Σ, the last equation can then be shortened to

$$\text{mean} = \sum_{\text{all } m} m \times P(M = m)$$

where $m = 1, 2, 4$ are the possible values taken by M.

The quantity 2 is the **mean** or **expectation** or **expected value** of the random variable M, written $E(M)$, in the example above.

In general, for a discrete random variable X, which can take specific values of x, the expected value (mean) of the random variable is defined by

$$E(X) = \sum_{\text{all } x} x \times P(X = x)$$

where the summation over 'all x' means all values of x for which the random variable X has a non-zero probability.

Example

When throwing a normal die, let X be the random variable defined by

$$X = \text{ the square of the score shown on the die.}$$

What is the expectation of X?

Solution

The possible values of X are

$$1,\ 2^2,\ 3^2,\ 4^2,\ 5^2 \text{ and } 6^2$$

$$\Rightarrow \quad 1,\ 4,\ 9,\ 16,\ 25 \text{ and } 36.$$

Each one has a probability of $\frac{1}{6}$ of occurring, so

$$E(X) = 1\times\frac{1}{6} + 4\times\frac{1}{6} + 9\times\frac{1}{6} + 16\times\frac{1}{6} + 25\times\frac{1}{6} + 36\times\frac{1}{6}$$

$$= \frac{1}{6}\times 91$$

$$= 15\frac{1}{6}.$$

Finally in this section, an alternative definition of a random variable will be developed.

In the previous example, what is the value of $\sum_{\text{all } x} P(X=x)$ **?**

If the summation is over all possible values of x, the summation must add up to one. So an alternative way of defining a discrete random variable is to impose the condition

$$\sum_{\text{all } x} P(X=x) = 1$$

Exercise 4A

1. Categorise each of the following as continuous or discrete. The random variables are:
   (a) A is 'the age in completed years of the first person I see wearing a hat'.
   (b) B is 'the length of the next car to enter the car park'.
   (c) C is 'how many cows I will see before the first green one'.
   (d) D is 'the date next July of the day with the highest temperature'.

2. Let $X =$ total score when two dice are rolled.
   (a) Find the possible values of the random variable X and determine the associated possibilities.
   (b) Determine the expectation of X.

   Could you answer (b) without actually performing any calculations?

3. Let $X =$ number of heads obtained when tossing a fair coin 3 times.
   (a) What are the possible values of X?
   (b) What are the associated probabilities?
   (c) Determine the mean value of X.

4. The random variable X has the probability distribution shown.

x	1	2	3	4	5
$P(X=x)$	$7c$	$5c$	$4c$	$3c$	c

   (a) Use the fact that $\Sigma P(X=x)=1$ to find c.
   (b) Explain why you expect $E(X)$ to be 3, greater than 3, or less than 3.
   (c) Calculate $E(X)$.

5. The random variable Z has probability distribution:

z	2	3	5	7	11
$P(Z=z)$	$\frac{1}{6}$	$\frac{1}{3}$	$\frac{1}{4}$	x	y

   and $E(Z) = 4\frac{2}{3}$. Find x and y.

4.2 Variance

The expression for the expected value just produced looks very similar to that in Chapter 3 which gave

$$\bar{x} = \frac{\Sigma x}{n}$$

for the mean value of a set of data.

A formula for variance like that from Chapter 3 can also be derived. Note that the variance was defined by

$$s^2 = \frac{\Sigma x^2}{n} - \bar{x}^2.$$

In the example in the previous section, groups of 36 terms could be expected, on average, to give a variance of

$$\frac{1^2 \times 12 + 2^2 \times 18 + 4^2 \times 6}{36} - 2^2$$

$$= 1^2 \times \frac{12}{36} + 2^2 \times \frac{18}{36} + 4^2 \times \frac{6}{36} - 2^2$$

$$= 1^2 P(M=1) + 2^2 P(M=2) + 4^2 P(M=4) - 2^2.$$

In general, the variance is defined by

$$\boxed{V(X) = E(X^2) - [E(X)]^2}$$

and the standard deviation, s, is as before defined by

$$s^2 = V(X).$$

For the example above,

$$V(M) = E(M^2) - [E(M)]^2$$

$$= \Sigma m^2 P(M=m) - [\Sigma m P(M=m)]^2,$$

giving $$V(M) = 1^2 \times \frac{1}{3} + 2^2 \times \frac{1}{2} + 4^2 \times \frac{1}{6} - 2^2$$

$$= \frac{1}{3} + 2 + \frac{8}{3} - 4$$

$$= 1.$$

As with data, the standard deviation gives a measure of the spread of the distribution.

Example

Find the variance and standard deviation of X, where

X = the square of the score shown on a die.

Solution

The possible values of X are

$$1, 4, 9, 16, 25 \text{ and } 36$$

each one having a probability of $\frac{1}{6}$. As you saw in Section 4.1, the mean value (expectation) is given by

$$E(X) = \tfrac{1}{6}(1 + 4 + 9 + 16 + 25 + 36) = 15\tfrac{1}{6}$$

whilst

$$E\left(X^2\right) = \tfrac{1}{6}\times 1 + \tfrac{1}{6}\times 16 + \tfrac{1}{6}\times 81 + \tfrac{1}{6}\times 256 + \tfrac{1}{6}\times 625 + \tfrac{1}{6}\times 1296$$

$$= \tfrac{1}{6}\times 2275$$

$$= 379\tfrac{1}{6}.$$

Thus $$V(X) = \frac{2275}{6} - \left(\frac{91}{6}\right)^2 = \frac{13650 - 8281}{36}$$

$$= \frac{5369}{36}$$

and $$s = \sqrt{\frac{5369}{36}} \approx 12.2.$$

Exercise 4B

1. If X is the score on a fair die, find the variance and standard deviation of X.
2. Find the variance and standard deviation of X when X is defined as in:

   (a) Exercise 4A, Question 2;

   (b) Exercise 4A, Question 3.
3. A team of 3 is to be chosen from 3 boys and 4 girls. If X is the random variable 'the number of girls in the team', find:

   (a) $E(X)$;

   (b) $V(X)$ and the standard deviation.

4.3 Probability density functions

The probabilities for the game in Activity 1 can be written in a table as:

m	1	2	4
$P(M = m)$	$\frac{1}{3}$	$\frac{1}{2}$	$\frac{1}{6}$

This gives the **probability distribution** of M as it shows how the total probability of 1 is distributed over the possible values. The function which assigns these probabilities is know as the **probability density function**, often abbreviated to p.d.f. and denoted by $p(m)$.

So $\quad p(1) = P(M=1) = \frac{1}{3}, \quad p(2) = \frac{1}{2}, \quad p(3) = \frac{1}{6}.$

In general, $P(X=x) = p(x)$, and the function p can often be written as a formula.

Example

The discrete random variable X has p.d.f.

$$p(x) = \frac{x}{36} \text{ for } x = 1, 2, 3, \ldots, 8.$$

Find $E(X)$ and $V(X)$.

Solution

Substituting the values 1 to 8 into the p.d.f. gives

x	1	2	3	4	5	6	7	8
$p(x)$	$\frac{1}{36}$	$\frac{2}{36}$	$\frac{3}{36}$	$\frac{4}{36}$	$\frac{5}{36}$	$\frac{6}{36}$	$\frac{7}{36}$	$\frac{8}{36}$

(The p.d.f. is a shorter way of giving all the probabilities associated with the random variable than drawing up a table, and indeed, there is no need to write one out if you do not feel it helps.)

As expected, note that the sum of all the probabilities is 1.

$$\begin{aligned} E(X) &= \Sigma x\, p(x) \\ &= 1 \times \frac{1}{36} + 2 \times \frac{2}{36} + 3 \times \frac{3}{36} + \ldots + 8 \times \frac{8}{36} \\ &= \frac{(1+4+9+\ldots+64)}{36} \\ &= \frac{204}{36} \\ &= 5\frac{2}{3}. \end{aligned}$$

Does this seem likely?

Well, the values five to eight have greater probabilities than one to four so the expected answer should be more than $4\frac{1}{2}$.

$$V(X) = \left(1^2 \times \frac{1}{36} + 2^2 \times \frac{2}{36} + \ldots + 8^2 \times \frac{8}{36}\right) - \left(5\frac{2}{3}\right)^2$$

$$= \frac{(1 + 8 + 27 + \ldots + 512)}{36} - 32\frac{1}{9}$$

$$= 36 - 32\frac{1}{9}$$

$$= 3\frac{8}{9}.$$

In the following section, you will consider some special p.d.f.'s which have wide applicability.

Exercise 4C

1. For a discrete random variable Y the p.d.f. is

   $$p(y) = \frac{5-y}{10} \quad \text{for } y = 1, 2, 3, 4.$$

   Calculate: (a) $E(Y)$ (b) $V(Y)$.

2. For a fair 10-sided spinner, if S is 'the score on the spinner', find:

   (a) the p.d.f. of S; (b) $E(S)$;

   (c) the standard deviation of S.

3. A random variable has p.d.f.

x	0	1	2	3
$P(X = x)$	0.4	0.3	0.2	0.1

   Find:

   (a) the mean and variance of X;

   (b) the mean and variance of the random variable

   $$Y = X^2 - 2X.$$

4. A fair six-sided die has

   '1' on one face
   '2' on two of its faces
   '3' on the remaining three faces.

   The die is thrown twice, and X is the random variable 'total score thrown'. Find

   (a) the p.d.f.;

   (b) the probability that the total score is more than 4;

   (c) $E(X)$ and $V(X)$.

4.4 The uniform distribution

One important distribution is the uniform one in which all possible outcomes have equal possibilities.

Activity 2 A survey of car registration plates

Survey vehicles in a car park or at any convenient place and note the digits on the number plates. Draw up a table like the one shown opposite.

What do you notice about the distribution of digits?

Digit	Tally	Frequency	Relative frequency
0			
1			
2			
.			
.			
.			
9			

The random variable X is said to follow a **uniform distribution** when all its outcomes are equally likely. A very simple example is given by the random variable H, 'the number of heads seen when a single coin is tossed'.

The probability density function is given by

h	0	1
$p(h)$	$\frac{1}{2}$	$\frac{1}{2}$

or $$p(h) = \frac{1}{2} \text{ for } h = 0, 1.$$

So $$E(H) = 0 \times \frac{1}{2} + 1 \times \frac{1}{2} = \frac{1}{2}$$

and $$V(H) = 0^2 \times \frac{1}{2} + 1^2 \times \frac{1}{2} - \left(\frac{1}{2}\right)^2$$

$$= \frac{1}{2} - \frac{1}{4} = \frac{1}{4}$$

and the standard deviation of H is $\frac{1}{2}$.

Activity 3 How random is your calculator?

Computers and certain calculators have a facility to enable you to generate random numbers. Some calculators will produce random numbers to three decimal places from 0.000 to 0.999. By simply reading only the first figure after the decimal point you can produce a set of random digits which should be uniformly distributed. Use your calculator to produce one hundred random digits and draw a frequency diagram to see how close your results are to the expected values.

Another example of a uniform distribution is the random variable, X, the score obtained when rolling a single unbiased die. In this case

$$P(X = x) = \frac{1}{6} \quad \text{for } x = 1, 2 \ldots, 6$$

giving $$E(X) = \frac{7}{2} = 3\frac{1}{2},$$

which can be written down by considering symmetry, and

$$V(X) = \frac{1}{6}(1 + 4 + 9 + 16 + 25 + 36) - \frac{49}{4}$$

$$= \frac{35}{12}.$$

This can be generalised to a random variable, X, having n equally likely outcomes for which the p.d.f. is given by

$$\boxed{P(X = x) = \frac{1}{n} \quad \text{for } x = 1, 2 \ldots, n}$$

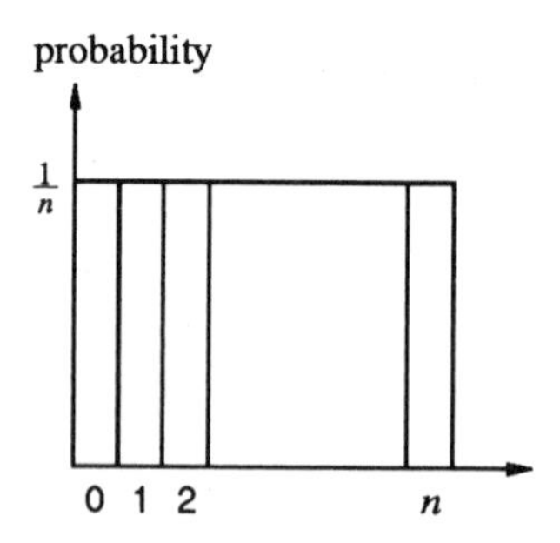

Activity 4

For a 30-sided spinner, let X be the score obtained. Determine

(a) $E(X)$ (b) $V(X)$.

You may be able to spot patterns in the results with n equal to 6 and 30 and deduce the general results.

i.e.

$n = 6$ $\quad E(X) = \frac{7}{2}$ $\quad V(X) = \frac{35}{12}$

$n = 30$ $\quad E(X) = \frac{31}{2}$ $\quad V(X) = \frac{899}{12}$

n $\quad E(X) = ?$ $\quad V(X) = ?$

The number 899 looks 'clumsy' but it can be written as

$900-1=30^2-1$, which suggests $V(X)=\dfrac{n^2-1}{12}$, although

$E(X)=\dfrac{n+1}{2}$ is simpler to see.

You will need the series summation results:

$$1+2+3+\ldots+n = \frac{n}{2}(n+1) \qquad (1)$$

and

$$1^2+2^2+3^2+\ldots+n^2 = \frac{n}{6}(n+1)(2n+1) \qquad (2)$$

to understand the proofs of the general results.

So
$$E(X) = 1\times\frac{1}{n} + 2\times\frac{1}{n} + 3\times\frac{1}{n} + \ldots + n\times\frac{1}{n}$$

$$= (1+2+3+\ldots+n)\times\frac{1}{n}, \quad \text{using (1),}$$

$$= \frac{n}{2}(n+1)\times\frac{1}{n}.$$

$$\boxed{E(X) = \frac{n+1}{2}}$$

Activity 5

Show, using equation (2), that

$$V(X) = \frac{n^2-1}{12}.$$

4.5 Miscellaneous Exercises

1. Find the probability density function for the random variable

   (a) the number of sixes obtained when two ordinary dice are thrown,

   (b) the smaller or equal number when two ordinary dice are thrown,

   (c) the number of heads when three fair coins are tossed.

2. For the discrete random variable X, the probability density function is given by

$$P(X=x)=\begin{cases} kx & x=1,2,3,4,5 \\ k(10-x) & x=6,7,8,9 \end{cases}$$

Find:

(a) the value of the constant k

(b) $E(x)$

(c) $V(x)$

3. Ten identically shaped discs are in a bag; two of them are black, the rest white. Discs are drawn at random from the bag in turn and not replaced.

Let X be the number of discs drawn up to and including the first black one.

List the possible values of X and the associated theoretical probabilities.

Calculate the mean value of X and its standard deviation. What is the most likely value of X?

If, instead, each disc is replaced before the next is drawn, construct a similar list of values and point out the chief differences between the two lists.

4. On a long train journey a statistician is invited by a gambler to play a dice game. The game uses two ordinary dice which the statistician is to throw. If the total score is 12, the statistician is paid £6 by the gambler. If the total score is 8, the statistician is paid £3 by the gambler. However, if both or either dice show a 1, the statistician pays the gambler £2. Let £X be the amount paid to the statistician by the gambler after the dice are thrown once.

Determine the probability that

(a) $X=6$, (b) $X=3$, (c) $X=-2$.

Find the expected value of X and show that, if the statistician played the game 100 times, his expected loss would be £2.78, to the nearest penny.

Find the amount, £a, that the £6 would have to be changed to in order to make the game unbiased. (SUJB)

5. A and B each roll a fair die simultaneously. Construct a table for the difference in their scores showing the associated probabilities. Calculate the mean of the distribution. If the difference in scores is 1 or 2, A wins; if it is 3, 4 or 5, B wins and if it is zero, they roll their dice again. The game ends when one of the players has won. Calculate the probability that A wins on (a) the first, (b) the second, (c) the rth roll. What is the probability that A wins?

If B stakes £1 what should A stake for the game to be fair? (SUJB)

6. A gambler has 4 packs of cards, each of which is well shuffled and has equal numbers of red, green and blue cards. For each turn he pays £2 and draws a card from each pack. He wins £3 if he gets 2 red cards, £5 if he gets 3 red cards and £10 if he gets 4 red cards.

(a) What are the probabilities of his drawing 0, 1, 2, 3, 4 red cards?

(b) What is the expectation of his winnings (to the nearest 10p)? (SUJB)

7. A player throws a die whose faces are numbered 1 to 6 inclusively. If the player obtains a six he throws the die a second time, and in this case his score is the sum of 6 and the second number; otherwise his score is the number obtained. The player has not more than two throws.

Let X be the random variable denoting the player's score. Write down the probability distribution of X, and determine the mean of X.

Show that the probability that the sum of two successive scores is 8 or more is $\frac{17}{36}$. Determine the probability that the first of two successive scores is 7 or more, given that their sum is 8 or more.

8. The discrete random variable X can take only the values 0, 1, 2, 3, 4, 5. The probability distribution of X is given by the following:

$$P(X=0)=P(X=1)=P(X=2)=a$$

$$P(X=3)=P(X=4)=P(X=5)=b$$

$$P(X\geq 2)=3P(X<2)$$

where a and b are constants.

(i) Determine the values of a and b.

(ii) Show that the expectation of X is $\frac{23}{8}$ and determine the variance of X.

(iii) Determine the probability that the sum of two independent observations from this distribution exceeds 7.

5 BINOMIAL DISTRIBUTION

Objectives

After studying this chapter you should

- be able to recognise when to use the binomial distribution;
- understand how to find the mean and variance of the distribution;
- be able to apply the binomial distribution to a variety of problems.

5.0 Introduction

'Bi' at the beginning of a word generally denotes the fact that the meaning involves 'two' and binomial is no exception. A random variable follows a binomial distribution when each trial has exactly **two possible outcomes**. For example, when Sarah, a practised archer, shoots an arrow at a target she either hits or misses each time. If X is 'the number of hits Sarah scores in ten shots', then the probabilities associated with $0, 1, 2, ..., 10$ hits can be expected to follow a particular pattern, known as the **binomial distribution**.

5.1 Finding the distribution

You have already met this type of distribution in Chapter 4, as can be seen in the following example.

Example

Ashoke, Theo and Sadie will each visit the local leisure centre to swim on one evening next week but have made no arrangement between themselves to meet or go on any particular day. The random variable X is 'the number of the three who go to the leisure centre on Wednesday'. Find the probability density function for X.

Solution

The probabilities of 0, 1, 2 or 3 people going on Wednesday can be found by using the tree diagram method covered in Section 1.5.

The following tree diagram shows probabilities for how many go on Wednesday.

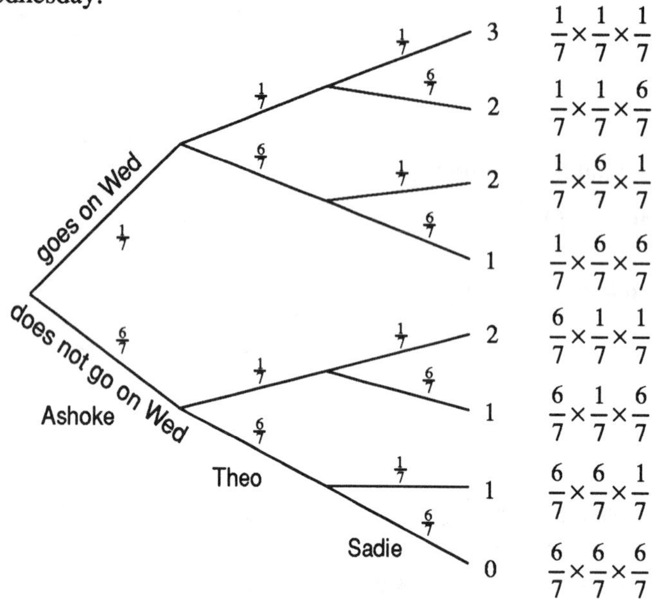

You can see that

$$P(X=3)=\left(\frac{1}{7}\right)^3, \quad P(X=0) = \left(\frac{6}{7}\right)^3$$

and $$P(X=2)=\frac{1}{7}\times\frac{1}{7}\times\frac{6}{7}+\frac{1}{7}\times\frac{6}{7}\times\frac{1}{7}+\frac{6}{7}\times\frac{1}{7}\times\frac{1}{7}$$

$$= 3\times\left(\frac{1}{7}\right)^2\left(\frac{6}{7}\right)$$

$$P(X=1)=\frac{1}{7}\times\frac{6}{7}\times\frac{6}{7}+\frac{6}{7}\times\frac{1}{7}\times\frac{6}{7}+\frac{6}{7}\times\frac{6}{7}\times\frac{1}{7}$$

$$= 3\times\left(\frac{1}{7}\right)\left(\frac{6}{7}\right)^2.$$

This gives the table below:

x	0	1	2	3
$P(X=x)$	$\left(\frac{6}{7}\right)^3$	$3\left(\frac{1}{7}\right)\left(\frac{6}{7}\right)^2$	$3\left(\frac{1}{7}\right)^2\left(\frac{6}{7}\right)$	$\left(\frac{1}{7}\right)^3$
	$=\frac{216}{343}$	$=\frac{108}{343}$	$=\frac{18}{343}$	$=\frac{1}{343}$

The method used in the example above can be extended to a fourth person so that there will be sixteen branches to cover all the possibilities as shown in the diagram below, with X now 'the number of the four people who go on Wednesday'.

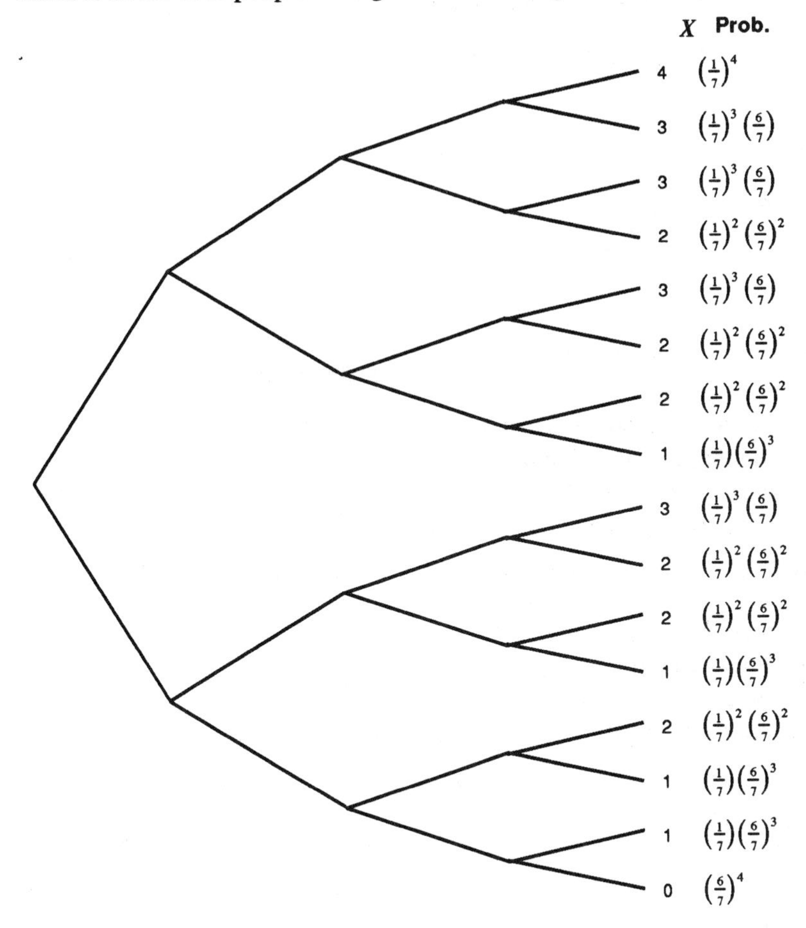

The resulting probability distribution is

x	0	1	2	3	4
$P(X=x)$	$\left(\frac{6}{7}\right)^4$	$4\left(\frac{1}{7}\right)\left(\frac{6}{7}\right)^3$	$6\left(\frac{1}{7}\right)^2\left(\frac{6}{7}\right)^2$	$4\left(\frac{1}{7}\right)^3\left(\frac{6}{7}\right)$	$\left(\frac{1}{7}\right)^4$

The fractions are as you might expect. For instance, looking at

$$P(X=2)=6\left(\frac{1}{7}\right)^2\left(\frac{6}{7}\right)^2,$$

since you are interested in having two present and the probability for each is $\frac{1}{7}$, the $(\frac{1}{7})^2$ is explained, and as two are not to attend this produces the $(\frac{6}{7})^2$.

Explain the reason for the coefficient 6.

This has come from the second of the tree diagrams: there are six branches corresponding to two being present. This is because there are six ways of writing down two $\left(\frac{1}{7}\right)$s and two $\left(\frac{6}{7}\right)$s in a row and each produces a branch.

The six ways are shown below:

$$\frac{1}{7}\times\frac{1}{7}\times\frac{6}{7}\times\frac{6}{7},\quad \frac{1}{7}\times\frac{6}{7}\times\frac{1}{7}\times\frac{6}{7},\quad \frac{1}{7}\times\frac{6}{7}\times\frac{6}{7}\times\frac{1}{7},$$

$$\frac{6}{7}\times\frac{1}{7}\times\frac{1}{7}\times\frac{6}{7},\quad \frac{6}{7}\times\frac{1}{7}\times\frac{6}{7}\times\frac{1}{7},\quad \frac{6}{7}\times\frac{6}{7}\times\frac{1}{7}\times\frac{1}{7}.$$

There are four fractions to write down, two of each type, and the number of different ways of combining these is six. You don't want to draw a tree diagram every time so another method can be developed. For example, having ten people going to the leisure centre would need a tree with $2^{10} = 1024$ branches.

If you can produce the fractions in the probability distribution then all that is needed is a way of getting the right numbers to put with them. With three people the numbers were 1, 3, 3, 1 and with four they were 1, 4, 6, 4, 1.

You might recognise these as being rows of **Pascal's Triangle**, shown opposite.

Pascal's Triangle

1 1

1 2 1

1 3 3 1

1 4 6 4 1

1 5 10 10 5 1

1 6 15 20 15 6 1

1 7 21 35 35 21 7 1

1 8 28 56 70 56 28 8 1

1 9 36 84 126 126 84 36 9 1

1 10 45 120 210 252 210 120 45 10 1

For three to be present you need $\left(\frac{1}{7}\right)^3$ and for seven to stay away the term is $\left(\frac{6}{7}\right)^7$. When writing down these fractions, how many different ways are there of combining them? According to Pascal's Triangle there are 120, so

$$P(X=3) = 120\left(\frac{1}{7}\right)^3\left(\frac{6}{7}\right)^7$$

which is easier to calculate than drawing 1024 branches! The number of ways of choosing 3 from 10 is often written

$$\binom{10}{3} \quad \text{or} \quad {}^{10}C_3,$$

so
$$p(X=3) = \binom{10}{3}\left(\frac{1}{7}\right)^3\left(\frac{6}{7}\right)^7$$

and the p.d.f. is

$$P(X=x) = \binom{10}{3}\left(\frac{1}{7}\right)^x\left(\frac{6}{7}\right)^{10-x} \qquad x = 0, 1, ..., 10.$$

Whilst the values needed can easily be read off Pascal's Triangle, there is an even easier way of working out the coefficients given in terms of factorials. Note that $n! = n(n-1)...2.1$ and, for example,

$$\binom{10}{3} = \frac{10!}{7!\,3!}$$

$$= \frac{10\times9\times8\times(7\times6\times5\times4\times3\times2\times1)}{(7\times6\times5\times4\times3\times2\times1)\times(3\times2\times1)}$$

$$= \frac{10\times9\times8}{3\times2\times1} \qquad \left(\text{since } \frac{10!}{7!} = 10\times9\times8\right)$$

$$= 120.$$

Activity 1

Check the values of

$$\binom{10}{1}, \binom{10}{2}, \binom{10}{4} \text{ and } \binom{10}{5}$$

by using Pascal's Triangle and the factorial formula.

It should also be noted that most calculators have the facility to produce the coefficients. Another notation used for the number of different ways of combining three from ten is ${}^{10}C_3$, and the majority of calculators have a key labelled with one of ${}^{n}C_r$ or $\binom{n}{r}$. Your instruction book will tell you how to use this function.

In general, when the probability of success is p (instead of $\frac{1}{7}$), and the experiment is repeated n independent times, the p.d.f. for the number of successes is given by

$$P(X=x) = \binom{n}{x}p^x(1-p)^{n-x} \qquad x = 0, 1, ...n$$

The notation $X \sim B(n, p)$ is often used.

Activity 2 E.S.P. test

Take five cards numbered from one to five. Seat two people back to back and give the cards to one of them. This person selects a card at random and the other participant tries to identify it.

This is done five times and repeated with other pairs of people. Record the result.

Now, if X is 'the number predicted correctly out of five attempts', the probability density function is

$$P(X=x) = \binom{5}{x}(0.2)^x(0.8)^{5-x} \quad \text{for } x = 0,\ 1,\ \dots\ 5.$$

So, for example, the probability of getting one correct is given by

$$P(X=1) = \binom{5}{1}(0.2)(0.8)^4$$

$$= 5(0.2)\,(0.8)^4$$

$$= 0.4096.$$

Hence for 20 people, you would expect

$$20 \times 0.4096 = 8.192 \approx 8$$

to get one correct. A table of expected frequencies is shown opposite.

Number correct (x)	$p(x)$	Expected frequencies $20\ p(x)$
0	0.3277	6.554
1	0.4096	8.192
2	0.2084	4.096
3	0.0512	1.024
4	0.0064	0.128
5	0.0032	0.064

Compare your observed frequencies from Activity 2.

How closely do they match the expected frequencies?

You may want to discard results from anyone who you feel has not co-operated, perhaps saying the same number every time.

Example

If X is binomially distributed with 6 trials and a probability of success equal to $\frac{1}{4}$ at each attempt, what is the probability of:

(a) exactly 4 successes (b) at least one success?

Solution

This question can be rewritten in the following way.

If $X \sim B\left(6, \frac{1}{4}\right)$, what is: (a) $P(X=4)$ (b) $P(X \geq 1)$?

(a) $$P(X=4) = \binom{6}{4}\left(\frac{1}{4}\right)^4\left(\frac{3}{4}\right)^2$$

$$= 15 \times \frac{1}{256} \times \frac{9}{16}$$

$$= \frac{135}{4096} = 0.033 \text{ (to 3 d.p.).}$$

(b) $$P(X \geq 1) = 1 - P(X=0)$$

$$= 1 - \left(\frac{3}{4}\right)^6$$

$$= 1 - \frac{729}{4096}$$

$$= \frac{3367}{4096} = 0.822 \text{ (to 3 d.p.).}$$

Note that tables giving cumulative binomial probabilities are given in the Appendix and these can be used where appropriate.

Example

When an unbiased coin is tossed eight times what is the probability of obtaining:

(a) less than 4 heads (b) more than five heads?

Solution

$H =$ number of heads $\Rightarrow H \sim B(8,\ 0.5)$

(a) Using the appropriate table in the Appendix you can simply write down the answer,

$$P(H<4) = P(H \leq 3)$$

$$= 0.3633.$$

Alternatively,

$$\begin{aligned}
P(H \le 3) &= P(H=0) + P(H=1) + P(H=2) + P(H=3) \\
&= \left(\frac{1}{2}\right)^8 + \binom{8}{1}\left(\frac{1}{2}\right)\left(\frac{1}{2}\right)^7 + \binom{8}{2}\left(\frac{1}{2}\right)^2\left(\frac{1}{2}\right)^6 \\
&\qquad + \binom{8}{3}\left(\frac{1}{2}\right)^3\left(\frac{1}{2}\right)^5 \\
&= \left(\frac{1}{2}\right)^8 + 8\left(\frac{1}{2}\right)^8 + 28\left(\frac{1}{2}\right)^8 + 56\left(\frac{1}{2}\right)^8 \\
&= 93\left(\frac{1}{2}\right)^8 \\
&= \frac{93}{256} \\
&= 0.3633 \text{ (to 4 d.p.).}
\end{aligned}$$

(b) $$\begin{aligned}
P(H > 5) &= 1 - P(H \le 5) \\
&= 1 - 0.8555 \quad \text{(from the table)} \\
&= 0.1445.
\end{aligned}$$

or

$$\begin{aligned}
P(H > 5) &= P(H=6) + P(H=7) + P(H=8) \\
&= \binom{8}{6}\left(\frac{1}{2}\right)^6\left(\frac{1}{2}\right)^2 + \binom{8}{7}\left(\frac{1}{2}\right)^7\left(\frac{1}{2}\right) + \left(\frac{1}{2}\right)^8 \\
&= 28\left(\frac{1}{2}\right)^8 + 8\left(\frac{1}{2}\right)^8 + \left(\frac{1}{2}\right)^8 \\
&= 37\left(\frac{1}{2}\right)^8 \\
&= \frac{37}{256} \\
&= 0.1445 \text{ (to 4 d.p.).}
\end{aligned}$$

Discuss why it is true that $\binom{8}{2}$ is the same as $\binom{8}{6}$.

Will it always be true that $\binom{n}{r} = \binom{n}{n-r}$?

Exercise 5A

Where decimals are used give answers correct to 3 significant figures.

1. If $X \sim B(6, \frac{1}{3})$ find:

   (a) $P(X=2)$ (b) $P(X<2)$ (c) $P(X \geq 1)$.

2. If $X \sim B(10, 0.3)$ find:

   (a) $P(X=9)$ (b) $P(X=0)$ (c) $P(X \leq 5)$.

3. A regular tetrahedron has three white faces and one red face. It is rolled four times and the colour of the bottom face is noted. What is the most likely number of times that the red face will end downwards?

4. If the probability that I get a lift to work on any morning is 0.6 what is the probability that in a working week of five days I will get a lift only twice?

5. When a consignment of pens arrives at the retailer's, ten of them are tested. The whole batch is returned to the wholesaler if more than one of those selected is found to be faulty. What is the probability that the consignment will be accepted if 2% of the pens are faulty?

5.2 The mean and variance of the binomial distribution

If you play ten games of table tennis against an opponent who, from past experience, you know only has a $\frac{1}{5}$ chance of winning a game with you, how many games do you expect him to win?

Most people would reply 'two' and would argue that since the opponent wins on average $\frac{1}{5}$ of the games he can expect to be successful in $\frac{1}{5} \times 10 = 2$.

Another way of writing this would be to say, if $X \sim B(10, \frac{1}{5})$, what is the value of $E(X)$? The answer then is $E(X) = 10 \times \frac{1}{5} = 2$.

In general, if $X \sim B(n, p)$, then the expected value of X is given by

$$E(X) = np$$

The formal proof of this result requires some work from pure maths, and in particular uses the result that

$$\binom{n}{r} = \frac{n!}{r!(n-r)!}.$$

Note that it will **not** be examined in the AEB Statistics paper.

$$\text{Now } E(X) = \sum_{0}^{n} x\,p(x)$$

$$= 0\binom{n}{0}q^n + 1\binom{n}{1}pq^{n-1} + 2\binom{n}{2}p^2q^{n-2}$$

$$+ 3\binom{n}{3}p^3q^{n-3} + ... + n\binom{n}{n}p^n, \text{ where } q = 1-p$$

$$= \frac{n!}{1!(n-1)!}pq^{n-1} + \frac{2\times n!}{2!(n-2!)}p^2q^{n-2}$$

$$+ \frac{3\times n!}{3!(n-3)!}p^3q^{n-3} + \ldots + np^n$$

$$= np\left\{\frac{(n-1)!}{1!(n-1)!}q^{n-1} + \frac{(n-1)!}{1!(n-2)!}pq^{n-2}\right.$$

$$\left. + \frac{(n-1)!}{2!(n-3)!}p^2q^{n-3} + \ldots + p^{n-1}\right\}$$

$$= np\left\{q^{n-1} + (n-1)q^{n-2}p + \frac{(n-1)(n-2)}{2}q^{n-3}p^2 + \right.$$

$$\left. \ldots + p^{n-1}\right\}$$

$$= np\,(p+q)^{n-1}, \text{ using the Binomial Theorem}$$

$$= np, \text{ since } p+q=1.$$

The variance of X is given by

$$\boxed{V(X) = npq}$$

The proof is even more complex than the analysis above, so it is set in the next Activity, which is optional.

*Activity 3

Using the formula

$$V(X) = E(X^2) - [E(X)]^2$$

show that

$$V(X) = npq.$$

---

*Activity 4

Use a computer package designed to show graphs of binomial distributions for different values of n and p to look at a variety of binomial distributions. In particular, identify the most likely outcome by picking out the tallest bar and see, for example, how much more spread out are the outcomes for $B(40, 0.5)$ than $B(10, 0.5)$.

Example

A biased die is thrown thirty times and the number of sixes seen is eight. If the die is thrown a further twelve times find:

(a) the probability that a six will occur exactly twice;

(b) the expected number of sixes;

(c) the variance of the number of sixes.

Solution

(a) Let X be defined by 'the number of sixes seen in twelve throws'.

Then $X \sim B(12, p)$ where $p = \frac{8}{30} = \frac{4}{15}$.

Since $X \sim B(12, \frac{4}{15})$,

$$P(X=2) = \binom{12}{2}\left(\frac{4}{15}\right)^2\left(\frac{11}{15}\right)^{10}$$

$$= \frac{66 \times 4^2 \times 11^{10}}{15^{12}}$$

$$= 0.211 \quad \text{(to 3 d.p.)}.$$

(b) $E(X) = np = 12 \times \frac{4}{15} = 3.2.$

(c) $V(X) = npq = 12 \times \frac{4}{15} \times \frac{11}{15} = 2\frac{26}{75} = 2.347$ (to 3 d.p.).

Example

A random variable X is binomially distributed with mean 6 and variance 4.2. Find $P(X \le 6)$.

Solution

Since X is a binomial distribution,

$$\text{mean} = np = 6,$$

$$\text{variance} = npq = 4.2.$$

Dividing, $q = \frac{4.2}{6} = 0.7$

and so $p = 1 - q = 0.3.$

This gives $0.3n = 6$

$\Rightarrow$ $n = 20.$

Hence $X \sim B(20, 0.3) \Rightarrow P(X \le 6) = 0.6080$ (from tables).

Activity 5 Binomial quiz

Ask your fellow students, and anyone else who will participate, whether the following statements are 'true' or 'false'.

1. *The Portrait of a Lady* was written by Henry James.
2. Psalms is the 20th book of the Old Testament.
3. The equatorial diameter of Mercury is about 3032 miles.
4. Mankoya is a place in Zambia.
5. 'The Potato Eaters' is a painting by Cezanne.
6. The Battle of Sowton was fought in 1461.

Make a frequency table to show the number of correct answers out of six for those asked.

Is there any evidence from your results that people really know some or all of the answers?

If they are just guessing, the number of correct answers, C say, should follow a binomial distribution,

$$C \sim B(6, \tfrac{1}{2}).$$

Work out $P(C=0)$, $P(C=1)$, ..., $P(C=6)$ and multiply them by the number of people asked to get the frequencies with which you would expect 0, 1,...., 6 correct answers to occur. Draw a diagram to show how your observed and expected frequencies compare.

Note that the tables in the Appendix contain pages of **cumulative** binomial probabilities. For example, $B\,(10,\ 0.1)$ has entries as shown opposite.

x	$P(X \le x)$
0	0.3487
1	0.7361
2	0.9298
3	0.9872
4	0.9984
5	0.9999
6	1.0000

So if X is distributed $B(10,0.1)$, then $P(X \le 3) = 0.9872$.

But
$$\begin{aligned} P(X=3) &= P(X \le 3) - P(X \le 2) \\ &= 0.9872 - 0.9298 \\ &= 0.0574. \end{aligned}$$

This figure may be inaccurate in the last digit as it has come from two numbers which have both been rounded. Using tables usually saves having to do several calculations and the benefit is considerable in cases such as $Y \sim B(30,\ 0.1)$ when the value of quantities like $P(Y \le 8)$ are needed.

Exercise 5B

1. On average a bowler takes a wicket every eight overs. What is the probability that he will bowl ten overs without succeeding in getting a wicket?

2. How many times must an unbiased coin be tossed so that the probability that at least one tail will occur is at least 0.99?

3. The random variable X has a binomial distribution $B(11,p)$. If $P(X=8)=P(X=7)$ find the value of P.

4. 100 families each with three children are found to have the following number of boys.

Number of boys	0	1	2	3
Frequency	13	34	40	13

   (a) Find the probability that a single baby born is a boy.

   (b) Calculate the number of families with three children you would expect to have two boys in a sample of 100 using your value from (a).

5. A multiple choice test has twenty questions and five possible answers for each one with only one correct per question. If X is 'the number of questions answered correctly' give:

   (a) the distribution of X;

   (b) the mean and variance of X;

   (c) the probability that a student will achieve a pass mark of 10 or more purely by guessing.

6. Investigate your results from the 'drawing pins' question in Exercise1B to see if X, the number of times a pin finishes point upwards in 10 trials, follows a binomial distribution.

7. The probability that a student will pass a maths test is 0.8. If eighteen students take the test, give the distribution of X, 'the number of students who pass', and find its most likely value.

8. (Drunkard's Walk) A drunk is ten steps away from falling in the dock. Every step he takes is either directly towards or away from the dock and he is equally likely to move in either direction. Find the probability that he will fall in the dock on his

(a) 10th step (b) 12th step.

Find also the probability that he is further from the dock after ten steps than he was at the start.

9. Find the probability that at most four heads will occur when a coin is tossed ten times.

10. If the probability that Don will hit a target on any shot is 0.2 and the probability for Yvette is 0.4, which of them is more likely to score at least three hits if Don has ten goes and Yvette has five goes?

5.3 Miscellaneous Exercises

1. (a) The probability that a certain type of vacuum tube will shatter during a thermal shock test is 0.15.

What is the probability that if 25 such tubes are tested

(i) 4 or more will shatter,

(ii) between 16 and 20 (inclusive) will survive?

Another type of tube is tested in samples of 30. It is observed that on 40% of occasions all 30 survive the test. What is the probability (assumed constant) of a single tube of this type surviving the test?

(b) A monkey in a cage is rewarded with food if it presses a button when a light flashes. Say, giving reasons, whether or not it is likely that the following variables follow the binomial distribution:

(i) Y is the number of times the light flashes before the monkey is twice successful in obtaining the food.

(ii) Z is the number of times that the monkey obtains the food by the time the light has flashed 20 times. (AEB)

2. A die is biased and the probability, p, of throwing a six is known to be less than $\frac{1}{6}$. An experiment consists of recording the number of sixes in 25 throws of the die. In a large number of experiments the standard deviation of the number of sixes is 1.5. Calculate the value of p and hence determine, to two places of decimals, the probability that exactly three sixes are recorded during a particular experiment.

3. (a) For each of the experiments described below, state, giving reasons, whether a binomial distribution is appropriate.

Experiment 1 A bag contains black, white and red marbles which are selected at random, one at a time with replacement. The colour of each marble is noted.

Experiment 2 This experiment is a repeat of Experiment 1 except that the bag contains black and white marbles only.

Experiment 3 This experiment is a repeat of Experiment 2 except that marbles are not replaced after selection.

(b) On average 20% of the bolts produced by a machine in a factory are faulty. Samples of 10 bolts are to be selected at random each day. Each bolt will be selected and replaced in the set of bolts which have been produced on that day.

(i) Calculate, to 2 significant figures, the probability that, in any one sample, two bolts or less will be faulty.

(ii) Find the expected value and the variance of the number of bolts in a sample which will not be faulty.

4. A crossword puzzle is published in *The Times* each day of the week, except Sunday. A man is able to complete, on average, 8 out of 10 of the crossword puzzles.

(a) Find the expected value and the standard deviation of the number of completed crosswords in a given week.

(b) Show that the probability that he will complete at least 5 in a given week is 0.655 (to 3 significant figures).

(c) Given that he completes the puzzle on Monday, find, to 3 significant figures, the probability that he will complete at least 4 in the rest of the week.

(d) Find, to 3 significant figures, the probability that, in a period of four weeks, he completes 4 or less in only one of the four weeks.

5. At a certain university in Cambford students attending a first course in statistics are asked by the lecturer, Professor Thomas Bayes, to complete 10 example sheets during the course. At the end of the course each student sits an examination as a result of which he either passes of fails. Assuming that

(i) the number, N, of example sheets completed by any student has a binomial distribution given by

$$P(N=n)={}^{10}C_n\left(\frac{2}{3}\right)^n\left(\frac{1}{3}\right)^{(10-n)}$$
$$n=0,1,\dots,10$$

and

(ii) the probability of a student passing the examination given that he completed n sheets during the course, is $n/10$,

(a) what is the (unconditional) probability that a student passes the examination?

(b) What is the probability that a student selected at random from the examination pass list had in fact completed four example sheets or less? (AEB)

*6 Thatcher's Pottery produces large batches of coffee mugs decorated with the faces of famous politicians. They are considering adopting one of the following sampling plans for batch inspection.

Method A (single sample plan) Select 10 mugs from the batch at random and accept the batch if there are 2 or less defectives, otherwise reject batch.

Method B (double sample plan) Select 5 mugs from the batch at random and accept the batch if there are no defectives, reject the batch if there are 2 or more defectives, otherwise select another 5 mugs at random. When the second sample is drawn count the number of defectives in the combined sample of 10 and accept the batch if the number of defectives is 2 or less, otherwise reject the batch.

(a) If the proportion of defectives in a batch is p, find, in terms of p, for each method in turn, the probability that the batch will be accepted.

(b) Evaluate both the above probabilities for $p=0.2$ and $p=0.5$.

(c) Hence, or otherwise, decide which of these two plans is more appropriate, and why. (AEB)

6 POISSON DISTRIBUTIONS

Objectives

After studying this chapter you should

- be able to recognise when to use the Poisson distribution;
- be able to apply the Poisson distribution to a variety of problems;
- be able to approximate the binomial distribution by a suitable Poisson distribution.

6.0 Introduction

This distribution is introduced through the Activity below.

Activity 1 Vehicle survey

The survey should be carried out on a motorway or dual carriageway, well away from any obstacles that would prevent free flow of traffic such as road works, roundabouts or traffic lights. (A useful, but not essential, piece of apparatus is something that will 'beep' at the end of each minute. Some watches with alarms can be set to go off every minute, or perhaps your chess club has a lightning buzzer. It is something that can be made without much difficulty by an electronics student.)

Note the number of vehicles that pass you in one minute and also the number of lorries. Collect readings for 100 minutes in a period which is either wholly in or completely outside any 'rush hour' surge of traffic that might exist.

For each of your distributions calculate:

(a) mean, $E(X)$ (b) variance, $V(X)$

where X is ' number of vehicles (or lorries) passing in one minute'.

(If several people are involved then the far carriageway can be studied separately. The 'number of red cars' is another example of a variable that could be examined.)

The Poisson distribution,which is developed in the next section, is of particular use when the number of possible occurrences of an event is unlimited. Possible examples are when describing the number of:

(a) flaws in a given length of material;

(b) accidents on a particular stretch of road in a week;

(c) telephone calls made to a switchboard in one day.

6.1 Developing the distribution

Surveys of the type undertaken in Activity 1 are important for transport planners who have to make decisions about road building schemes. For example, in considering the need for an extra lane on a dual carriageway, a survey of the number of lorries over several days gave the number passing a point per minute during the evening rush hour, as in the table opposite.

No. of lorries per minute (x)	Frequency (f)
0	7
1	34
2	84
3	140
4	176
5	176
6	146
7	104
8	65
9	36
10	18
11	8
12	4
13	1
14	1
≥15	0

Here is an example of a random variable X, 'the number of lorries per minute', which is certainly going to produce a discrete probability distribution, but each one minute trial will have many possible outcomes.

This cannot be a binomial distribution since, in theory at least, the possible values of X are unlimited.

Now the mean of the data is given by

$$\bar{x} = \frac{0\times7 + 1\times34 + ... + 14\times1}{1000} \approx 4.997$$

whilst the variance is

$$s^2 = \frac{0^2\times7 + 1^2\times34 + ... + 14^2\times1}{1000} - 4.997^2 \approx 5.013.$$

So allowing a little for experimental error, it seems that the distribution has its mean **equal** to its variance. A relationship between succeeding frequencies can be seen by dividing consecutive data.

$$\frac{34}{7} \approx \frac{5}{1} \quad \frac{84}{34} \approx \frac{5}{2} \quad \frac{140}{84} = \frac{5}{3} \quad \frac{176}{140} \approx \frac{5}{4}$$

$$\frac{176}{176} = \frac{5}{5} \quad \frac{146}{176} \approx \frac{5}{6} \quad \frac{104}{146} \approx \frac{5}{7} \quad \frac{65}{104} = \frac{5}{8}$$

$$\frac{36}{65} \approx \frac{5}{9} \quad \frac{18}{36} = \frac{5}{10}$$

The initial probability, $P(X=0) = 0.007$, can then be used to calculate the others.

$$P(X=1) = \frac{5}{1}P(X=0)$$

$$P(X=2) = \frac{5}{2}P(X=1) = \frac{5^2}{2\times1}P(X=0)$$

$$P(X=3) = \frac{5}{3}P(X=2) = \frac{5^3}{3\times2\times1}P(X=0)$$

$$P(X=4) = \frac{5^4}{4\times3\times2\times1}P(X=0).$$

Hence the p.d.f. can be written

$$P(X=n) = \frac{5^n}{n(n-1)\;....\;2\times1}P(X=0)$$

$$= \frac{5^n}{n!}P(X=0), \quad \text{using factorials.}$$

Since the sum of the probabilities is one, putting $p = P(X=0)$, it follows that

$$1 \;=\; p + 5p + \frac{5^2 p}{2!} + \frac{5^3 p}{3!} + \frac{5^4 p}{4!} + ...$$

$$= p\left(1 + 5 + \frac{5^2}{2!} + \frac{5^3}{3!} + \frac{5^4}{4!} + ...\right)$$

$$= pe^5, \text{ since } e^5 = 1 + 5 + \frac{5^2}{2!} + ...,$$

$$\Rightarrow \quad p = e^{-5}.$$

The exponential number $e \approx 2.71828$ is a very important number in advanced mathematical analysis and can be found on all scientific calculators. The exponential function, e^x, takes the form

$$e^x = 1 + x + \frac{x^2}{2!} + \frac{x^3}{3!} + ...$$

and it is this result that has been used above with $x=5$.

Activity 2 Telephone calls

Many school and college switchboards are computerised and can provide a print-out of calls over any period. Other switchboards in local firms may also be able to help. Study the number of incoming calls in, for example, ten minute periods, during a time of day avoiding lunch and other breaks. Look at the results for several days. Calculate the mean and variance of your distribution and try to fit a Poisson distribution to your figures.

Activity 3

As an alternative or additional practical to Activity 2, study the number of arrivals of customers at a post office in two minute intervals. The length of the time interval may well be shortened in the case of a large and busy site.

The key parameter in fitting a **Poisson distribution** is the mean value, usually denoted by λ. This is the average number of occurrences in the specified period (e.g. cars passing in a minute). The probability distribution function is then given by:

x	0	1	2	3	4	...
$P(X=x)$	$e^{-\lambda}$	$\lambda e^{-\lambda}$	$\frac{\lambda^2 e^{-\lambda}}{2!}$	$\frac{\lambda^3 e^{-\lambda}}{3!}$	$\frac{\lambda^4 e^{-\lambda}}{4!}$	...

In general, if X is a Poisson distribution, then

$$P(X=x) = \frac{\lambda^x e^{-\lambda}}{x!} \quad (x = 0, 1, 2, \ldots)$$

and this is denoted by $X \sim Po(x)$.

The **Poisson distribution** was first derived in 1837 by the French mathematician *Simeon Denis Poisson* whose main work was on the mathematical theory of electricity and magnetism.

The distribution arises when the events being counted occur

(a) independently;

(b) such that the probability that two or more events occur simultaneously is zero;

(c) randomly in time or space;

(d) uniformly (that is, the mean number of events in an interval is directly proportional to the length of the interval).

Example

If the random variable X follows a Poisson distribution with mean 3.4, find $P(X = 6)$.

Solution

This can be written more quickly as: if $X \sim Po(3.4)$ find $P(X = 6)$.

Now $$P(X=6) = \frac{e^{-\lambda}\lambda^6}{6!}$$

$$= \frac{e^{-3.4}(3.4)^6}{6!} \quad (\text{mean, } \lambda = 3.4)$$

$$= 0.071\,604\,409 = 0.072 \quad \text{(to 3 d.p.).}$$

Example

The number of industrial injuries per working week in a particular factory is known to follow a Poisson distribution with mean 0.5.

Find the probability that

(a) in a particular week there will be:

  (i) less than 2 accidents,

  (ii) more than 2 accidents;

(b) in a three week period there will be no accidents.

Solution

Let A be 'the number of accidents in one week'.

(a) (i) $$P(A<2) = P(A \le 1)$$

$$= 0.9098 \quad \text{(from tables, to 4 d.p.)}$$

or, from the formula,

$$P(A<2) = P(A=0) + P(A=1)$$

$$= e^{-0.5} + \frac{e^{-0.5} \times 0.5}{1!}$$

$$= \frac{3}{2} e^{-0.5}$$

$$\approx 0.9098.$$

(ii) $P(A > 2) = 1 - P(A \le 2)$

$$= 1 - 0.9856 \quad \text{(from tables)}$$

$$= 0.0144 \quad \text{(to 4 d. p.)}$$

or

$$1 - [P(A=0) + P(A=1) + P(A=2)]$$

$$= 1 - \left[e^{-0.5} + e^{0.5} 0.5 + \frac{e^{0.5}(0.5)^2}{2!} \right]$$

$$= 1 - e^{-0.5}(1 + 0.5 + 0.125)$$

$$= 1 - 1.625\, e^{-0.5}$$

$$\approx 0.0144.$$

(b) $P\,(0 \text{ in } 3 \text{ weeks}) = \left(e^{-0.5}\right)^3 \approx 0.223.$

Could the number of vehicles on a single carriageway road passing a fixed point in some time interval be expected to follow a Poisson distribution?

Exercise 6A

Give answers to 3 significant figures.

1. If $X \sim Po(3)$, find:

   (a) $P(X=2)$ (b) $P(X=3)$

   (c) $P(X \ge 5)$ (d) $P(X < 3)$.

2. If $X \sim Po(\lambda)$ and $P(X=4) = 3P(X=3)$, find λ and $P(X=5)$.

3. If $X \sim Po(\lambda)$ and $P(X=0) = 0.323$, find the value of λ to two decimal places and use this to calculate $P(X=3)$.

4. Investigate whether or not the following figures might result from a Poisson variable.

x	0	1	2	3	4	5	6	≥ 7
$P(X=x)$	0.368	0.368	0.184	0.061	0.015	0.003	0.001	0.000

*5. If $X \sim Po(2)$, $Y \sim Po(3)$ and $Z \sim Po(5)$, find:

   (a) $P(X+Y=0)$ (b) $P(X+Y=1)$ (c) $P(Z=0)$

   (d) $P(Z=1)$ (e) $P(X+Y \le 2)$ (f) $P(Z \le 2)$

6.2 Combining Poisson variables

Activity 4

The number of telephone calls made by the male and female sections of the P.E. department were noted for fifty days and the results are shown in the table opposite. The number of calls by men are given first in each pair of numbers.

Investigate the distributions of the numbers of calls made:

(a) by the male staff;

(b) by the female staff;

(c) in total each day i.e. $0+2=2$, $2+2=4$, etc.

0, 2	2, 2	6, 0	3, 5	1, 2
2, 2	1, 1	2, 2	1, 1	2, 3
7, 0	1, 4	3, 6	2, 3	3, 0
4, 1	5, 1	4, 3	5, 4	6, 4
1, 0	2, 3	3, 2	3, 3	6, 1
2, 3	2, 2	2, 1	3, 5	5, 3
4, 3	4, 2	3, 4	4, 3	3, 1
3, 1	3, 3	4, 4	5, 4	2, 1
5, 6	1, 2	2, 2	1, 2	3, 3
4, 2	0, 5	4, 4	2, 2	4, 1

Number of telephone calls (male, female)

Activity 5

Study the numbers of cars and lorries from your survey in Activity 1and look at the distributions of the numbers of:

(a) cars (b) lorries (c) all vehicles.

Now consider the result of combining two independent Poisson variables

$$A \sim Po(2) \text{ and } B \sim Po(3).$$

Define the new distribution $C = A + B$.

What can you say about C ?

You know that

$$P(A=0) = e^{-2}, \quad P(A=1) = e^{-2} \times 2, \quad P(A=2) = \frac{e^{-2}2^2}{2!}, \quad \ldots$$

$$P(B=0) = e^{-3}, \quad P(B=1) = e^{-3} \times 3, \quad P(B=2) = \frac{e^{-3}3^2}{2!}, \quad \ldots$$

This gives

$$P(C=0) = P(A=0) \times P(B=0) = e^{-2} \times e^{-3} = e^{-5}.$$

$$P(C=1) = P(A=0)\times P(B=1) + P(A=1)\times P(B=0)$$
$$= e^{-2}\times 3e^{-3} + 2e^{-2}\times e^{-3} = 5e^{-5},$$

and
$$P(C=2) = P(A=0)\times P(B=2) + P(A=1)\times P(B=1) + P(A=2)\times P(B=0)$$
$$= e^{-2}\times\frac{9e^{-3}}{2!} + 2e^{-2}\times 3e^{-3} + e^{-2}\frac{2^2}{2!}\times e^{-3}$$
$$= e^{-5}\left(\frac{9}{2!} + 6 + \frac{4}{2!}\right)$$
$$= \frac{25e^{-5}}{2!} = \frac{e^{-5}5^2}{2!}.$$

What sort of distribution do these results indicate?

In general, if $A \sim Po(a)$ and $B \sim Po(b)$ are independent random variables, then $C = (A+B) \sim Po(a+b)$.

You have seen this result illustrated in a special case above. The proof requires a good working knowledge of the binomial expansion and is set as an optional activity below.

*Activity 6

By noting that

$$P(C=n) = \sum_{i=0}^{n} P(A=i)\times P(B=n-i)$$

and that
$$(a+b)^n = \sum_{i=0}^{n}\binom{n}{i}a^i b^{n-i}$$

prove that $C \sim Po(a+b)$.

Example

The number of misprints on a page of the *Daily Mercury* has a Poisson distribution with mean 1.2. Find the probability that the number of errors

(a) on page four is 2; (b) on page three is less than 3;

(c) on the first ten pages totals 5;

(d) on all forty pages adds up to at least 3.

Solution

Let E be 'the number of errors on one page', so that
$E \sim Po(1.2)$.

(a) $P(E=2) = \dfrac{e^{-1.2}(1.2)^2}{2!} \approx 0.217.$

(b) $P(E<3) = P(E \le 2) \approx 0.8795$, from tables,

or

$$P(E<3) = P(E=0) + P(E=1) + P(E=2)$$

$$= e^{-1.2} + 1.2e^{-1.2} + \frac{(1.2)^2 e^{-1.2}}{2!}$$

$$= e^{-1.2}(1 + 1.2 + 0.72)$$

$$= 2.92e^{-1.2}$$

$$= 0.8795 \quad \text{(to 4 d.p.)}.$$

(c) Let E_{10} be 'the number of errors on 10 pages',

then $E_{10} \sim Po(12)$, as $E_{10} = E + E + \ldots + E$,

and $E_{10} \sim Po(1.2 + 1.2 + \ldots. + 1.2) = Po(12)$.

Hence $P(E_{10} = 5) = \dfrac{e^{-12}12^5}{5!} \approx 0.0127.$

(d) Similarly $E_{40} \sim Po(48)$.

$$P(E_{40} \ge 3) = 1 - P(E_{40} \le 2)$$

$$= 1 - \left(e^{-48} + e^{-48} \times 48 + \frac{e^{-48} \times 48^2}{2!}\right)$$

$$= 1 - 1201e^{-48} \approx 1.000 \text{ (to 3 d.p.)}.$$

Tables may well be a time saving device, as was true with the binomial distribution.

Activity 7

A firm has three telephone numbers. They all receive numbers of calls that follow Poisson distributions, the first having a mean of 8, the second 4 and the third 3 in a period of half an hour. Find the probability that

(a) the second and third lines will receive a total of exactly six calls in half an hour;

(b) the firm will receive at least twelve calls in half an hour;

(c) line two will receive at most six calls in one hour;

(d) line one will receive no calls in 15 minutes.

Example

A shop sells a particular make of video recorder.

(a) Assuming that the weekly demand for the video recorder is a Poisson variable with mean 3, find the probability that the shop sells

(i) at least 3 in a week,

(ii) at most 7 in a week,

(iii) more than 20 in a month (4 weeks).

Stocks are replenished only at the beginning of each month.

(b) Find the minimum number that should be in stock at the beginning of a month so that the shop can be at least 95% sure of being able to meet the demands during the month.

Solution

(a) Let X be the demand in a particular week. Thus $X \sim Po(3)$ and, using the Poisson tables in the Appendix,

(i) $P(X \geq 3) = 1 - P(X \leq 2)$

$= 1 - 0.4232$ (Note that the tables give the

$= 0.5768$ cumulative probabilities)

(ii) $P(X \leq 7) = 0.9881$, from tables.

(iii) If Y denotes the demand in a particular month, then

$Y \sim Po(12)$ and

$$P(Y > 20) = 1 - P(Y \leq 20)$$
$$= 1 - 0.9884$$
$$= 0.0116.$$

(b) You need to find the smallest value of n such that

$$P(Y \le n) \ge 0.95.$$

From the tables,

$$P(Y \le 17) = 0.9370, \quad P(Y \le 18) = 0.9626$$

So the required minimum stock is 18.

A very important property of the Poisson distribution is that if $X \sim Po(\lambda)$, then

$$E(X) = V(X) = \lambda.$$

That is, both the mean and variance of a Poisson distribution are equal to λ.

To show that $E(X) = \lambda$, note that, by definition,

$$E(X) = \sum_{\text{all } x} xP(X = x)$$

$$= 0 \times e^{-\lambda} + 1 \times \left(\lambda e^{-\lambda}\right) + 2 \times \left(\frac{\lambda^2 e^{-\lambda}}{2!}\right) + 3 \times \left(\frac{\lambda^3 e^{-\lambda}}{3!}\right) + \ldots$$

$$= \lambda e^{-\lambda}\left(1 + \lambda + \frac{\lambda^2}{2!} + \frac{\lambda^3}{3!} + \ldots\right)$$

$$= \lambda e^{-\lambda} e^{\lambda} \text{ (using the result from page 117)}$$

$$= \lambda$$

which proves the result. The proof of $V(X) = \lambda$ follows in a similar but more complicated way.

***Activity 8**

If $X \sim Po(\lambda)$ show that $V(X) = \lambda$.

Exercise 6B

1. Incoming telephone calls to a school arrive at random times. The average rate will vary according to the day of the week. On Monday mornings, in term time there is a constant average rate of 4 per hour. What is the probability of receiving

   (a) 6 or more calls in a particular hour,

   (b) 3 or fewer calls in a particular period of two hours?

   During term time on Friday afternoons the average rate is also constant and it is observed that the probability of no calls being received during a particular hour is 0.202. What is the average rate of calls on Friday afternoons?

   (AEB)

2. Write down two conditions which need to be satisfied in order to use the Poisson distribution.

   The demand at a garage for replacement windscreens occurs randomly and at an average rate of 5 per week.

   Determine the probability that no more than 7 windscreens are demanded in a week.

   The windscreen manufacturer uses glass which contains random flaws at an average rate of 48 per 100 m^2. A windscreen of area 0.95 m^2 is chosen at random.

   Determine the probability that the windscreen has fewer than 2 flaws.

   A random sample of 5 such windscreens is taken. Find the probability that exactly 3 of them contain fewer than 2 flaws. (AEB)

3. A garage uses a particular spare part at an average rate of 5 per week. Assuming that usage of this spare part follows a Poisson distribution, find the probability that

   (a) exactly 5 are used in a particular week,

   (b) at least 5 are used in a particular week,

   (c) exactly 15 are used in a 3-week period,

   (d) at least 15 are used in a 3-week period,

   (e) exactly 5 are used in each of 3 successive weeks.

   If stocks are replenished weekly, determine the number of spare parts which should be in stock at the beginning of each week to ensure that on average the stock will be insufficient on no more than one week in a 52 week year.

   (AEB)

4. A shopkeeper hires vacuum cleaners to the general public at £5 per day. The mean daily demand is 2.6.

   (a) Calculate the expected daily income from this activity assuming an unlimited number of vacuum cleaners is available.

   (b) Find the probability that the demand on a particular day is:

   (i) 0 (ii) exactly one

   (iii) exactly two (iv) three or more.

   (c) If only 3 vacuum cleaners are available for hire calculate the mean of the daily income.

   A nearby large store is willing to lend vacuum cleaners at short notice to the shopkeeper, so that in practice she will always be able to meet demand. The store would charge £2 per day for this service regardless of how many, if any, cleaners are actually borrowed. Would you advise the shopkeeper to take up this offer? Explain your answer. (AEB)

6.3 The Poisson distribution as an approximation to the binomial

Despite having tables and powerful calculators, it is often difficult to make binomial calculations if n, the number of experiments, becomes very large. In these circumstances, it is easier to approximate the binomial by a Poisson distribution.

Activity 9 Birthday dates

Obtain the dates of birth of the students in your college or school from official records or by running a survey. Record the number of people having a birthday on each date (omitting 29th February). Draw up a frequency table.

Use the theoretical probability $\frac{1}{365}$ to work out the expected frequencies of 0, 1, 2, ... people from the n people considered, who have birthdays on the same date, according to the binomial distribution $B\left(n, \frac{1}{365}\right)$. Using a mean number of people per day, $\frac{n}{365}$, calculate the expected frequencies from the Poisson distribution $Po\left(\frac{n}{365}\right)$.

Compare the observed and two sets of expected frequencies.

Poisson's first work on the distribution that bears his name arose from considering the binomial distribution and it was derived as an approximation to the already known binomial model.

When might a Poisson distribution give probabilities close to those of a binomial distribution?

If $X \sim B(n, p) \approx Po(\lambda)$, then the means and variances of the two distributions must be about the same. This gives

$$\text{mean} \quad : \quad np = \lambda$$

$$\text{variance} \quad : \quad npq = \lambda .$$

So, if p is small (for example $p = \frac{1}{365}$ in the Activity above),

then

$$q \approx 1 \text{ and } n = \frac{\lambda}{p}.$$

So when n is very large and p is very small, binomial probabilities may be approximated by Poisson probabilities with $\lambda = np$.

Normally, for this approximation it is required that

$$\boxed{n \geq 50 \text{ and } p \leq 0.1}$$

The approximation improves as $n \to \infty$, $p \to 0$.

Example

A factory produces nails and packs them in boxes of 200. If the probability that a nail is substandard is 0.006, find the probability that a box selected at random contains at most two nails which are substandard.

Solution

If X is 'the number of substandard nails in a box of 200', then

$$X \sim B(200,\ 0.006).$$

Since n is large and p is small, the Poisson approximation can be used. The appropriate value of λ is given by

$$\lambda = np = 200 \times 0.006 = 1.2.$$

So $\quad X \sim Po(1.2)$,

and $\quad P(X \le 2) = 0.8795$ (from tables),

or
$$P(X \le 2) = e^{-1.2} + e^{-1.2} \times 1.2 + \frac{e^{-1.2}1.2^2}{2!}$$
$$= 2.92e^{-1.2}$$
$$= 0.8795 \text{ (to 4 d.p.).}$$

One of the advantages of the use of a Poisson approximation is that tables can be used more often to avoid routine calculations.

Exercise 6C

Where appropriate, give answers to 3 significant figures.

1. If $X \sim B(500,\ 0.002)$, use the binomial and Poisson distributions to find:

   (a) $P(X=0)$ (b) $P(X=1)$ (c) $P(X=4)$.

2. If $X \sim B(200,\ 0.06)$, use Poisson tables to find the values of:

   (a) $P(X<20)$ (b) $P(X \ge 5)$.

3. Fuses are packed in boxes of 1000. If 0.2% are faulty find the probability that a box will contain

   (a) exactly 2 faulty; (b) at least one faulty.

4. A link in a metal chain has probability 0.03 of breaking under a load of 50 kg. What is the probability that a chain made of 100 such links will break when subjected to a 50 kg load?

5. The number of runs scored by Ali in an innings of a cricket match is distributed according to a Poisson distribution with mean 4.5. Find the probability that he will score:

   (a) exactly 4 in his next innings;

   (b) at least three in his next innings;

   (c) at least six in total in his next two innings.

6. State two conditions under which a binomial distribution may be approximated by a Poisson distribution, and give a reason why this approximation may be useful in practice.

   In the treatment of hay fever, the probability that any sufferer is allergic to a particular drug is 0.0005. Assuming that the occurrences of the allergy in different sufferers are independent, find the probability that in a random sample of 8000 sufferers more than four will be allergic to the drug.

   Each sufferer who is allergic to the drug has a probability of 0.3 of developing serious complications following its administration.

   (a) Determine the probability that, of the 8000 sufferers who are administered the drug, exactly two develop serious complications.

   In fact four sufferers develop the allergy following the administration of the drug to the random sample of 8000 sufferers.

   (b) Determine the probability that exactly two of these four develop serious complications.

   Explain, briefly, why your answers to (a) and (b) differ. (AEB)

6.4 Miscellaneous Exercises

1. The number of goals scored in a hockey match by Sarindar and Paula are independent Poisson variables with means 2.5 and 1.5 respectively. Find the probabilities that in a particular match:

   (a) Sarindar will score at least twice;

   (b) they will score at most three between them.

2. If 8% of a city is affected by an outbreak of flu, use the Poisson approximation to the binomial distribution to find the probability that a factory with 160 employees will have at least five people absent.

3. If X is a random variable which has a Poisson distribution with mean 3.99, what is the most likely value of X?

   Explain your result in terms of the relationship between frequencies.

4. A van hire firm has twelve vehicles available and has found that demand follows a Poisson distribution with mean 9.5. In a month of 25 working days, on how many days would you expect:

   (a) demand to exceed supply;

   (b) all vehicles to be idle;

   (c) it to be possible to service 3 of the vans?

5. The number of errors made by a typist on a single page is a Poisson variable with mean 0.09. Find the probability that a fifty page article will have:

   (a) at least 3 errors;

   (b) no errors on the first ten pages.

6. Use the Poisson approximation to the binomial distribution to calculate the probability that a consignment of 10 000 electronic components, each of which has a 0.02% probability of being faulty, contains only perfect items.

   If eight consignments are received, what is the most likely number to contain no faulty components?

7. If X is a random variable with p.d.f.

   $$p(x) = \frac{x}{c}, \text{ for } x = 4, 5, 6, 7, 8,$$

   find: (a) $E(X)$ (b) $V(X)$.

8. What is the probability that a Poisson variable with mean 5 will produce exactly two 3's in four trials?

9. The random variable X has probability distribution:

x	1	2	3
$p(x)$	a	0.5	b

   If the expected value of X is 1.7, find a and b and also $V(X)$.

10. A binomial distribution with N trials has a probability of success equal to seven times the probability of failure at each trial. If the probability of seven successes is four times that of six successes, find the value of N.

11. A newsagent finds that the mean number of copies of a particular magazine he sells each week is 10. If the number sold follows a Poisson distribution, find the probability that he sells less than four in a week.

    How many should he have in stock at the start of the week if the chance that he cannot provide a customer with a copy is less than 0.05?

12. Take any English novel and use random numbers to select a page and then a line on a page. Starting at the beginning of the line, count 50 letters and note the number of occurrences of double letters.

    Repeat the process until you have at least fifty results. Compare your figures with those from a Poisson distribution.

13. The number of bacteria in one millilitre of a liquid is known to follow a Poisson distribution with mean 3. Find the probability that a 1 ml sample will contain no bacteria. If 100 samples are taken, find the probability that at most ten will contain no bacteria. (Use a Poisson approximation and give your answer to the first part correct to 3 d.p.)

14. The numbers in a group booking into a hotel are found to follow a Poisson distribution with mean 2.2. What is the probability that the next booking will be for a party of more than three?

    What is the probability that just one of the next four bookings will be for such a group?

15. The number of parasites on fish hatched in the same season and living in the same pond follows a Poisson distribution with mean 3.6.

    Find, giving your answers to 3 decimal places, the probability that a fish selected at random will have

    (a) 4 or less parasites,

    (b) exactly 2 parasites. (AEB)

16. Customers at a motorway service station enter the cafeteria through a turnstile. The cafeteria is open 24 hours a day and an automatic counting device records the number of people entering each minute. State, giving reasons, whether or not it is likely that these data will follow a Poisson distribution. (AEB)

17. In the manufacture of commercial carpet, small faults occur at random in the carpet at an average rate of 0.95 per 20 m^2.

    Find the probability that in a randomly selected 20 m^2 area of this carpet

    (a) there are no faults

    (b) there at most 2 faults.

    The ground floor of a new office block has 10 rooms. Each room has an area of 80 m^2 and has been carpeted using the same commercial carpet described above.

    For any one of these rooms, determine the probability that the carpet in that room

    (c) contains at least 2 faults,

    (d) contains exactly 3 faults,

    (e) contains at most 5 faults.

    Find the probability that in exactly half of these 10 rooms the carpets will contain exactly 3 faults. (AEB)

18. A polytechnic offers a short course on advanced statistical methods. As the course involves a large amount of practical work only 8 places are available. Advertising starts two months before the course and if, at the end of one month, 3 or fewer places have been taken the course is cancelled. If 4 or more places have been taken by the end of one month the course is run regardless of the number of applications received in the second month. If the number of applications per month follows a Poisson distribution with mean 3.6, and places are allocated on a first come first served basis, what is the probability that at the end of one month the course will be

    (a) cancelled,

    (b) full?

    What is the probability that

    (c) a place will be available at the start of the second month,

    (d) the course will run with 8 students?

    If the course is offered on four separate occasions, what is the probability that it will

    (e) run once and be cancelled three times,

    (f) run with 8 students on 2 or more occasions?

    (AEB)

7 CONTINUOUS PROBABILITY DISTRIBUTIONS

Objectives

After studying this chapter you should

- understand the use of continuous probability distributions and the use of area to calculate probabilities;
- be able to use probability functions to calculate probabilities and find measures such as the mean and variance;
- recognise and be able to use the rectangular distribution.

7.0 Introduction

Note that in order to work through this chapter you will need to be able to

(a) factorise and expand polynomials up to order 3;

(b) integrate simple functions and use definite integrals to find areas under curves;

(c) differentiate simple functions and find turning points.

On virtually every food item purchased you will find a **nominal** weight. If you find a packet of crisps which weighs 24 g and the nominal weight is 25 g, are you entitled to complain? Clearly manufacturers cannot be expected to make every packet exactly 25 g but the law requires a certain percentage of all packets to be above this weight. The manufacturer therefore needs to know the pattern or distribution of the weights of the crisp packets in order to check whether or not the company is breaking the law.

Activity 1 Coin tossing

For this you will need a number of 2p coins. Mark out a playing grid with pieces of string on a tarmac area or on short grass, as shown opposite.

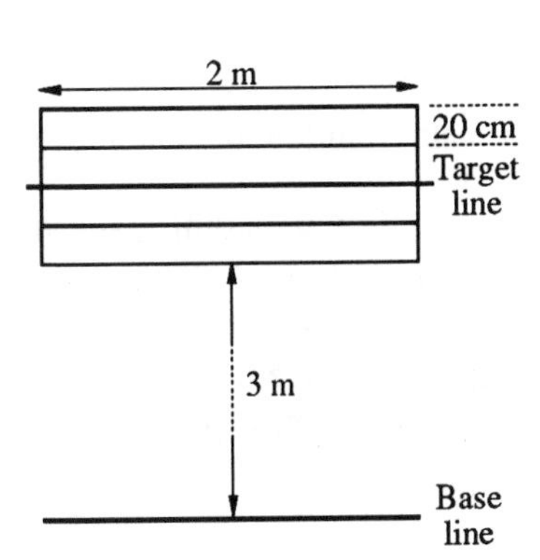

The central target line could be drawn in a different colour. The aim of the game is for a person standing on the base line to toss the coins to land as close to the target line as possible. Any coins falling outside the grid area are taken again. Let each member of the group try it a number of times to give about 100 results. Record the distance each coin lands away from the target line in centimetres noting whether it is in front or behind with – / + respectively.

If you wanted to write a computer program to 'simulate' this game you would need to know the probability of the coin landing different distances from the line. Provided your aim is fairly good you should expect to get more shots nearer the line than further away. The same idea is used in evaluating how likely an artillery weapon is to hit its target. Gunners can only estimate the distance to a target and shots will fall in a particular pattern around the target.

The main aim of this chapter is to develop a method of representing the probabilities in terms of a continuous function. This will enable estimates to be made as to what proportion will be within specified limits.

7.1 Looking at the data

In a similar experiment shot-putters were asked to aim at a line 10 m away. They threw the shot 200 times and throws were measured within 2 m either side of the line. The results were as shown below.

in front

metres	1.99-1.50	1.49-1.00	0.99-0.5	0.49-0
frequency	9	22	31	37

behind

metres	0-0.49	0.50-0.99	1.00-1.49	1.50-1.99
frequency	38	32	23	8

A histogram to represent these data is shown opposite.

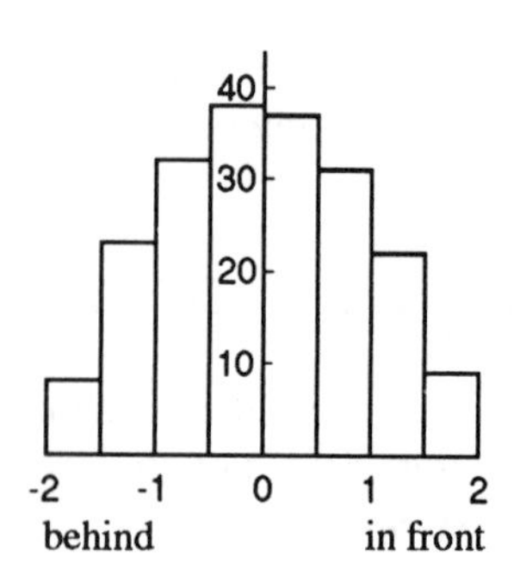

You could say that this pattern was the one that all throws were likely to follow. To obtain a more accurate picture it might be possible to collect more data; this would also allow narrower groups to be used. A possible pattern which might emerge is shown on the next page.

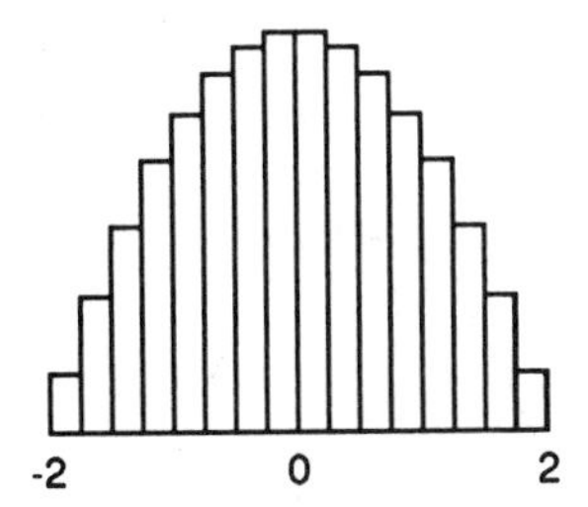

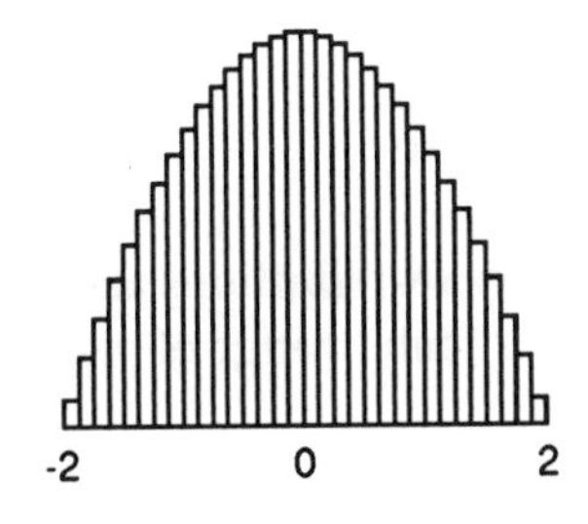

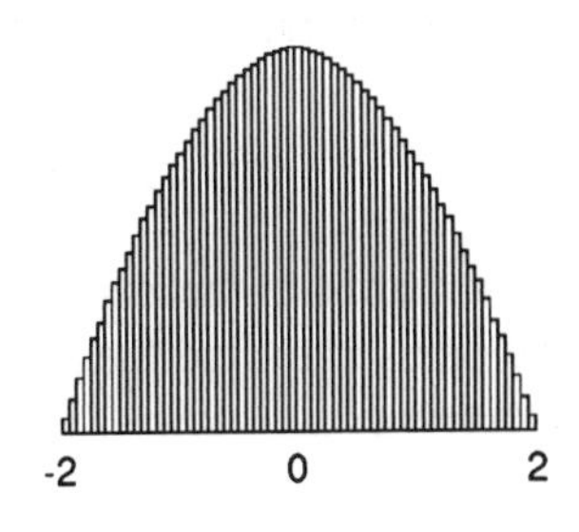

In the final case the bars are so thin it appears that the tops are a continuous curve - this is called a **frequency curve**. For very large samples then, the graph can be shown as opposite.

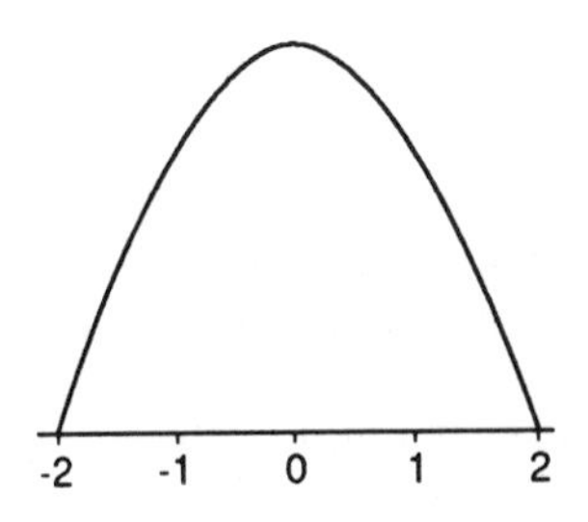

Unless you know the total sample size though, you cannot put a scale on the y-axis. So what should you use?

In how many throws out of 100 roughly would a shot putter throw a distance of exactly 10 m?

How many throws in 100 would you expect to be between 9.9 m and 10.1 m?

How many throws in 100 would be behind the target line?

How many would be between 1 m and 2 m beyond the line?

How would the answer differ for 500 throws or 5000?

Two factors should emerge from this, namely that

(a) the proportion of throws getting a precise value is infinitely small;

(b) for a particular range of values the **proportion** remains constant.

In order to answer some of the above questions more accurately you will need to measure the areas under the graph. From the original data the probability of any throw being between 0 and 1 m behind the line is

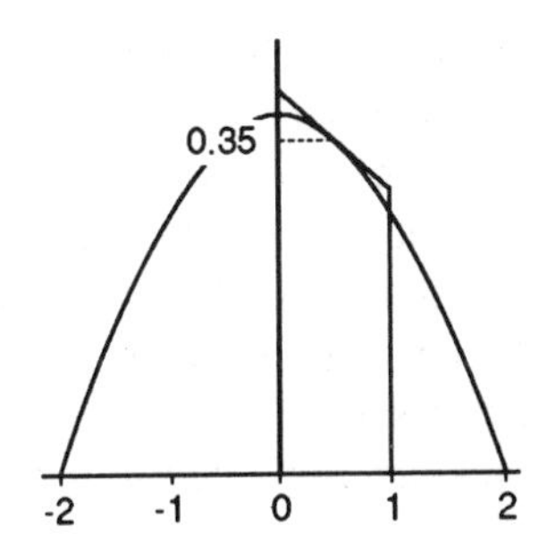

$$\frac{(38+32)}{200} = 0.35$$

On the graph we can estimate this by using a trapezium, which will have mid-point height 0.35 to give an area 0.35.

Using all the other results a scale can be given to the vertical axis. Note that, as shown opposite,

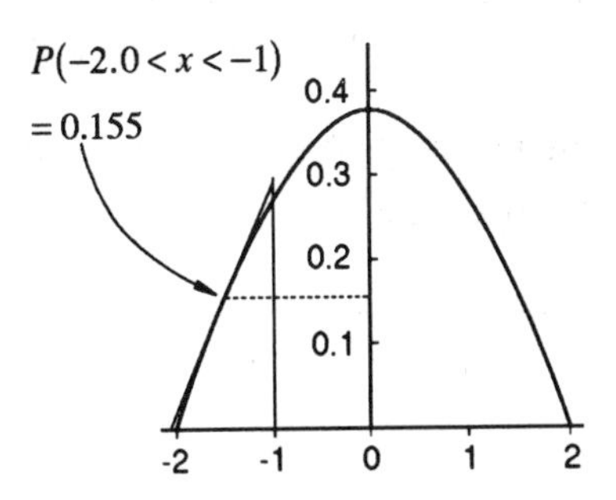

$$P(-2.0 < x < -1.0) = 0.155.$$

For smaller ranges the area principle still works; for example

$$P(0 < x < 0.5) = 0.5 \times 0.38 = 0.19.$$

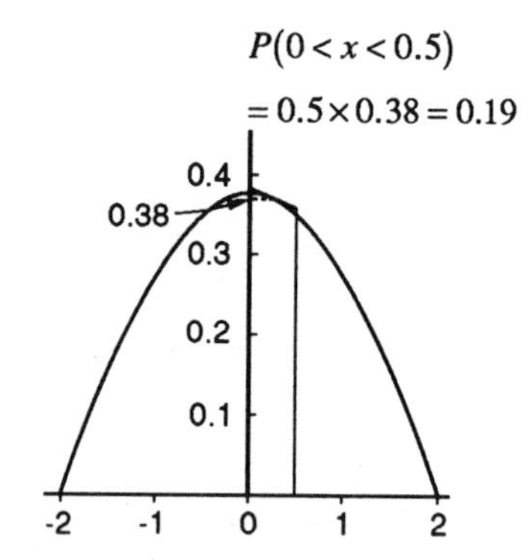

Such graphs as these are called **probability distributions** and they can be used to find the probability of a particular **range** of values occurring.

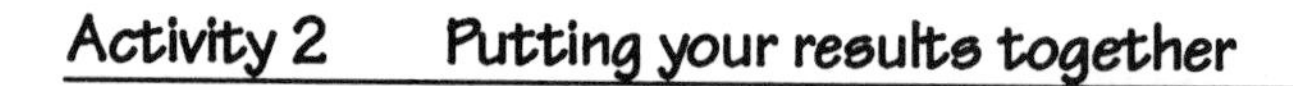

Activity 2 Putting your results together

Draw a histogram of your coin tossing results using 10 cm intervals. Draw a frequency curve of your results and scale with relative probabilities. Remember that for a 10 cm group the percentage frequencies will have to be divided by 10. Use your graph and trapeziums to find approximately:

(a) the probability of a coin landing between 10 and 30 cm past the line;

(b) the probability of a throw being more than 20 cm past the line;

(c) the probability of a throw being more than 15 cm short of the line.

7.2 Finding a function

Activity 3 Fishing for a function

The data on the page opposite represent the fish caught in one trawl by a scientific research vessel. These are all shown at 1/10th scale. A scientist wants to set up a probability distribution for the lengths of fishes so that she can simulate the catches in future, rather than take more fish. Measure the lengths of all the fish on the sheet to the nearest mm and record your results in a group table using a group size of 0.5 cm. A computer with a statistical or spreadsheet package might be useful in this exercise.

Draw a histogram and a probability distribution curve. Remember that percentage frequencies will need to be divided by the group size (i.e. doubled) for the vertical axis.

What is the probability of catching a fish bigger than 48 cm i.e. $P(x > 4.8)$?

1
2
3
4
5
6
7
8
9
10
11
12
13
14
15
16
17
18
19
20
21
22
23
24
25
26
27
28
29
30
31
32
33
34
35
36
37
38
39
40
41
42
43
44
45
46
47
48
49
50
51
52
53
54
55
56
57

Using a graph is clearly an inefficient and tedious way of calculating probabilities. You should be familiar with the technique of integration, in particular using it to find areas under curves. In order to use this concept, however, it is necessary to find the function to represent the curve.

If you look again at the curve for the shot putt throws on page 132, this appears to be an inverted quadratic function. Since it crosses the x-axis at ± 2, these are roots of the equation. The function is therefore of the form

$$f(x) = A(4 - x^2)$$

where A is constant.

To find the value of A it is necessary to make sure that the total area under the curve between -2 and $+2$ is 1, since the total probability must be one. This can be found by integration as follows:

Note that questions involving integration will not be set for the AEB examination of Statistics module, so much of this chapter is optional.

$$\begin{aligned}\int_{-2}^{2} A(4 - x^2)\,dx &= A\left[4x - \frac{x^3}{3}\right]_{-2}^{+2} \\ &= A\left\{\left[8 - \frac{8}{3}\right] - \left[-8 - \frac{-8}{3}\right]\right\} \\ &= \frac{32A}{3}.\end{aligned}$$

This must equal 1, so A must take the value $\frac{3}{32}$.

Note that this only works if the range of answers is restricted to -2 to $+2$. This is usually made clear by defining a probability distribution function (p.d.f.) as follows:

$$f(x) = \begin{cases} \frac{3}{32}(4 - x^2) & \text{for } -2 < x < 2 \\ 0 & \text{otherwise} \end{cases}$$

Any function which can be used to describe a continuous probability distribution is called a **probability distribution function.**

*Activity 4 Checking out functions

The scientist in the fish example wants to find a suitable function for her results. What kind of function do you think applies here? Use a graphic calculator or computer package to guess at possible alternatives.

For each of the alternatives below and any of your own, do the following:

(a) Find any constants that are necessary to give a total area of 1 under the curve (assume that all fish are between 0.5 and 5.5).

(b) Sketch the actual curve using a graphic calculator, computer package or table of values over the original histogram. Judge how well the curve fits.

A : quadratic curve
[Hint: assume it goes through (6, 0)]

B : straight line

C : cosine curve (use radians).

Care must be taken with some functions as, although they may have an area of 1, they may not be functions usable in calculating probability. For example, consider the function illustrated opposite.

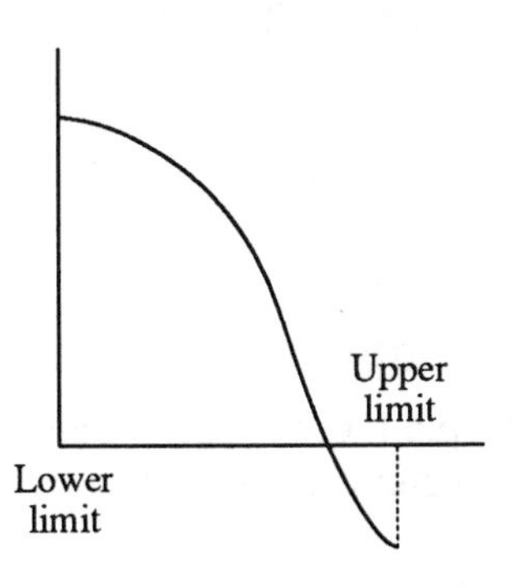

Although this function may integrate to give an area of 1, this is because the area below the graph is negative. This is **not** a probability distribution function, as it would give negative probabilities.

The only other restriction on suitable functions is that the function exists for all values of x in the given range. The best way to check this is to sketch the graph or plot on a graphic calculator.

*Exercise 7A

1. Check whether the following are suitable probability distribution functions over the given range:

(a) $f(x)=\frac{1}{2}\left(x^2+4\right) \quad 0<x<1$

(b) $f(x)=\frac{1}{2} \quad 2<x<4$

(c) $f(x)=\frac{x}{4} \quad 1<x<3$

(d) $f(x)=\frac{x}{6}+\frac{1}{12} \quad 0<x<3$

(e) $f(x)=\frac{1}{2}(2x-1) \quad 0<x<2$

2. A variable has a p.d.f. given by

$$f(x)=A\left(x^2+4\right) \quad 0<x<1.$$

Find the value of the constant A such that this constitutes a valid p.d.f.

7.3 Calculating probabilities

Once a suitable function has been found, the main purpose of using a p.d.f., that is, to calculate the probabilities of events, can be carried out. In the previous two sections two ideas have been used, namely

(a) that the probability of a range of values can be found by finding the area between under a p.d.f curve;

(b) that integration of a p.d.f. can be used to find these areas.

In general, if $f(x)$ is a continuous p.d.f. defined over a specified range of x, then

$$\text{total area under the curve} = 1 \Rightarrow \int_{-\infty}^{\infty} f(x)dx = 1$$

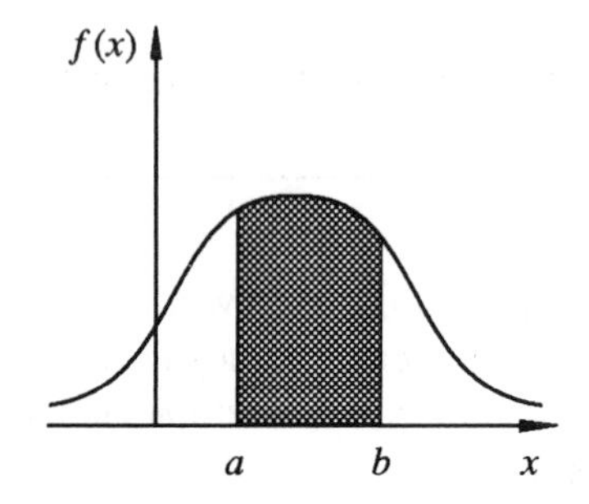

and $P(a < x < b)$ is the area under the curve from $x = a$ to $x = b$; this can be written as

$$P(a < x < b) = \int_a^b f(x)dx$$

*Example

The p.d.f. of the age of babies, x years, being brought to a post-natal clinic is given by

$$f(x) = \begin{cases} \frac{3}{4}x(2-x) & 0 < x < 2 \\ 0 & \text{otherwise} \end{cases}$$

If 60 babies are brought in on a particular day, how many are expected to be under 8 months old?

Solution

Eight months $= \frac{2}{3}$ year, so

$$P\left(x < \tfrac{2}{3}\right) = \int_0^{\frac{2}{3}} \tfrac{3}{4}x(2-x)dx$$

$$= \tfrac{3}{4}\int_0^{\frac{2}{3}} \left(2x - x^2\right)dx$$

$$= \tfrac{3}{4}\left(x^2 - \frac{x^3}{3}\right)_0^{\frac{2}{3}}$$

$$= \tfrac{3}{4}\left[\tfrac{4}{9} - \tfrac{8}{81}\right] - [0]$$

$$= \tfrac{3}{4}\left[\tfrac{28}{81}\right] = \tfrac{7}{27} \approx 0.259$$

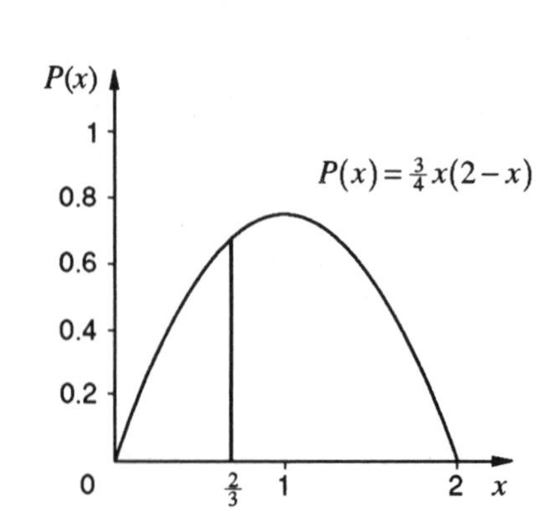

Hence the expected number of babies under 8 months

$$= 60 \times 0.259 = 15.5.$$

The function which was obtained after integration, and is used to calculate probabilities, is called the **cumulative distribution function**, $F(x)$. In the last example

$$F(x) = \frac{3}{4}\left(x^2 - \frac{x^3}{3}\right), \quad 0 \le x \le 2$$

It is cumulative, as putting a single value a in the function will give you the probability $P(0 < x < a)$.

For example,

P(baby is less than 6 months old)

$$= F(0.5) = \frac{3}{4}\left(0.5^2 - \frac{(0.5)^3}{3}\right) \approx 0.16$$

Once the cumulative function is known it is unnecessary to repeat the integration for new examples. In general, for a continuous p.d.f., $f(x)$, the cumulative distribution function, is given by

$$\boxed{F(x) = \int_{-\infty}^{x} f(x)dx}$$

*Activity 5 How good are your functions?

In the last activity different functions were suggested to fit the fish data. How can you measure which is the best? For each function find the probability of a value being in each 0.5 cm group. If you find $F(x)$ this should not take long. By multiplying each of these by 57 you can find the 'expected' number in each category according to the function. Make a table showing the expected and observed numbers.

How can an overall measure be found to test which is the best fit?

*Exercise 7B

You may assume, unless stated, that all the functions are valid p.d.f.s. Sketch the function in each case.

1. The resistance of an electrical component follows a p.d.f. given by

$$f(x) = \begin{cases} \dfrac{x}{4} & 1 < x < 3 \\ 0 & \text{otherwise} \end{cases}$$

What is the probability that the resistance is less than 2?

2. A biologist is examining the growth of a virus. A tiny amount is placed on a culture plate and it is found that the surface area in cm^2 occupied by the virus eight hours later is given by the p.d.f.

$$f(x) = \frac{e^x}{19} \quad 0 < x < 3.$$

Check that this is approximately a valid p.d.f. and find the probability that a culture plate has a surface area in excess of 2 cm^2.

3. The weekly demand for petrol at a local garage (in thousands of litres) is given by the p.d.f.

$$f(x) = 48\left(x - \tfrac{1}{2}\right)(1 - x) \quad \tfrac{1}{2} < x < 1.$$

The petrol tanks are filled to capacity of 940 litres every Monday. What is the probability the garage runs out of petrol in a particular week?

*7.4 Mean and variance

Activity 6 Response times

Using a stopwatch or a watch with a stopwatch facility in hundredths of a second, set the watch going and try to stop at exactly 5 seconds. Record the exact time on the stopwatch - again a computer facility would help. It is also easier to work in pairs. Repeat this 100 times and draw up a histogram of your results. Find also the mean and variance of your results.

Although it will vary according to how good you are, the following p.d.f. should approximate to your times.

$$f(x) = \begin{cases} \dfrac{375}{32}(5.4 - x)(x - 4.6) & 4.6 < x < 5.4 \\ 0 & \text{otherwise} \end{cases}$$

Sketch this curve and verify that it approximates to your data. Find the expected number of results in each range using the p.d.f. You could then work out the expected mean and variance using these figures, but is there a quicker way?

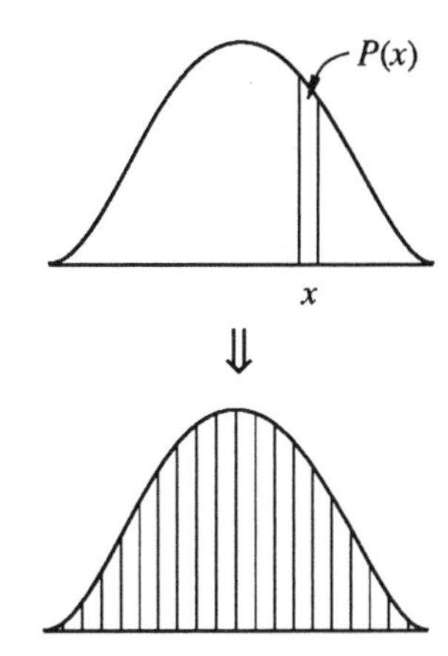

The idea of the mean and variance of a p.d.f. was met in the last chapter on discrete distribution. The mean can be found by the same method i.e. $E(X) = \Sigma x\, P(x)$. This would require multiplying small ranges of x by the area under the curve for the whole defined range.

Clearly this is tedious and to obtain an accurate result would require very small ranges of x. In the last section, integration was used to find areas over a range of values. To find the **mean** of a distribution simply use

$$E(X) = \int_{-\infty}^{\infty} x f(x)\, dx$$

In most cases the function is defined over only a small range so it is not necessary to integrate between $\pm\infty$, only in the defined range. This can be used in the baby example to find the mean age of babies brought to the clinic.

Example

Find the mean age of babies brought to the clinic described in the example in Section 7.3.

Solution

$$\begin{aligned}
E(X) &= \int_0^2 x \,.\, \frac{3}{4} x \,.\, (2 - x)\, dx \\
&= \frac{3}{4} \int_0^2 \left(2x^2 - x^3\right) dx \\
&= \frac{3}{4} \left[\frac{2}{3} x^3 - \frac{x^4}{4} \right]_0^2 \\
&= \frac{3}{4} \left[\frac{16}{3} - 4 \right] - [0] \\
&= \frac{3}{4} \times \frac{4}{3} = 1.
\end{aligned}$$

This result should not be surprising since the original sketch showed the distribution to be symmetrical, so the mean must be in the middle of the range. This will always be true, so this could save integrating.

In the same way, the basic definition of **variance** used with discrete distributions can be used, but replacing summation with integration; this gives

$$V(X) = E\left(X^2\right) - \mu^2$$

Example

For the babies distribution, find the variance of x.

Solution

$$E(X^2) = \int_0^2 x^2 \frac{3}{4}x(2-x)\,dx$$

$$= \frac{3}{4}\int_0^2 \left(2x^3 - x^4\right)dx$$

$$= \frac{3}{4}\left[\frac{1}{2}x^4 - \frac{1}{5}x^5\right]_0^2$$

$$= \frac{3}{4}\left[8 - \frac{32}{5}\right] - [0]$$

$$= \frac{3}{4} \times \frac{8}{5} = \frac{6}{5}.$$

Hence $$V(X) = \frac{6}{5} - 1^2 = \frac{1}{5}.$$

The standard deviation can be found by square-rooting the variance, so for the example above,

$$s = \sqrt{\frac{1}{5}} \approx 0.45.$$

Activity 7

Evaluate the mean and variance of the p.d.f. used in the response times activity. How well do these compare with the actual values?

*Exercise 7C

1. A p.d.f. is given by $f(x) = 6x(1-x)$ for $0 < x < 1$. Find the mean and variance of this distribution.

2. A teacher asks her pupils to draw a circle with some compasses they have been given. The p.d.f. of the radii is given by

   $$f(r) = \frac{r}{4} \quad 1 < r < 3.$$

   Find the mean and variance of the radii drawn.

3. The proportion of cloud cover at a particular meteorological office is given by the p.d.f.

   $$f(x) = 12x(1-x)^2 \quad 0 < x < 1.$$

   Find the mean and variance of this distribution.

4. A p.d.f. is given by $f(x) = ke^{-x}$ for $x > 0$. Find the value of k which makes this valid and hence the mean and variance of this distribution.

5. A p.d.f. is given by $f(x) = Ax(6-x)^2$ for $0 < x < 6$.

   Find the value of A and hence the mean and variance of this distribution.

*7.5 Modes, medians, quartiles

There are, of course, other measures which can be calculated for a p.d.f. Some of these are introduced below.

Mode

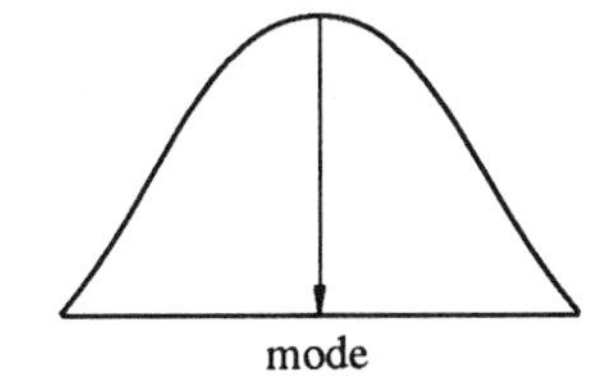

The mode is defined as the value which has highest frequency. In a continuous case this is clearly the value of x which gives the maximum value of the function.

In many cases a simple curve sketch will show this value.

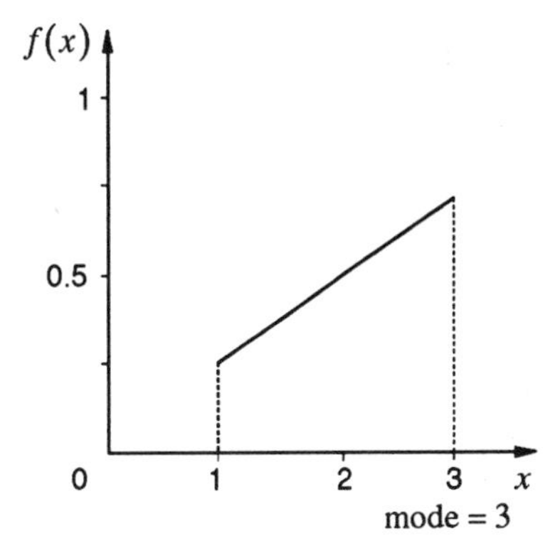

For example, when

$$f(x) = \frac{x}{4} \qquad 1 < x < 3,$$

the mode is clearly at 3.

With more complicated functions it may be necessary to differentiate to find **maxima/minima**.

Example

Find the mode of the p.d.f. defined by

$$f(x) = 12x^2(1-x) \qquad 0 < x < 1.$$

Solution

Since
$$f(x) = 12x^2 - 12x^3$$

$$\frac{d}{dx}f(x) = 24x - 36x^2$$

$$= 12x(2-3x)$$

$$= 0$$

when $x = 0$ or $\frac{2}{3}$.

A sketch (or second derivative) would reveal that $x = 0$ is a minimum point and $\frac{2}{3}$ is the mode.

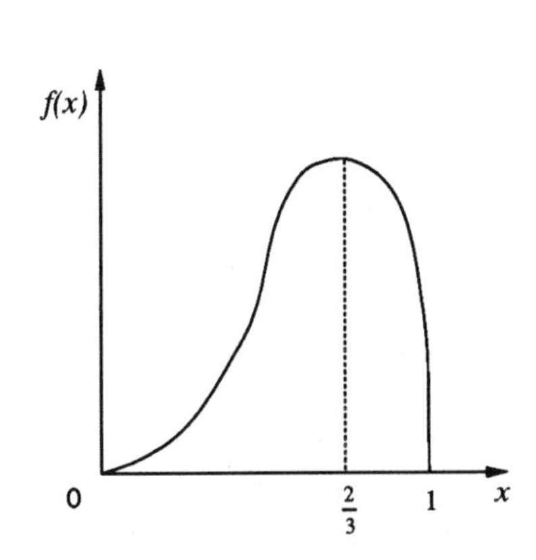

Activity 8

Look at the graphs of p.d.f.s that you drew in the fish example and in some of the exercises. For which cases is the mode obvious and which will require a maxima/minima differentiation method? Are there any for which it would be impossible to find a mode? Sketch some cases which might cause problems.

Median and quartiles

Medians and, in fact, any of the percentiles can be found from their basic definitions. The **median** is defined as the value m for which

$$P(x < m) = P(x > m) = \frac{1}{2}.$$

In terms of a continuous p.d.f. this is the value which divides the area into two parts each with area $\frac{1}{2}$.

The value m can be found by integration.

Example

Find the median of this p.d.f.

$$f(x) = \frac{x}{4}, \quad 1 < x < 3.$$

Solution

Now $\quad \int_1^m \frac{x}{4} dx = \frac{1}{2}$

$$\Rightarrow \quad \left[\frac{x^2}{8}\right]_1^m = \frac{1}{2}$$

$$\Rightarrow \quad \frac{m^2}{8} - \frac{1}{8} = \frac{1}{2}$$

$$\Rightarrow \quad m^2 - 1 = 4$$

$$\Rightarrow \quad m = \pm\sqrt{5}.$$

Since $-\sqrt{5}k$ is outside the range of the function the median must be at $\sqrt{5}$.

The same method can be used to find quartiles and other percentiles. However, for higher order polynomial equations solutions can be difficult. It is often simpler to use the cumulative distribution function.

Example

For the babies' weight example used earlier, the cumulative distribution function was given by

$$F(x) = \frac{3}{4}\left(x^2 - \frac{x^3}{3}\right), \quad 0 < x < 2.$$

Show that the median value is $x = 1$, and estimate the interquartile range.

Solution

Now $$F(1) = \frac{3}{4}\left(1 - \frac{1}{3}\right) = \frac{3}{4} \times \frac{2}{3} = \frac{1}{2}.$$

So the area from $x = 0$ to $x = 1$ is 0.5, and $x = 1$ is the median (as well as the mode and mean!).

For the lower quartile, $F(x) = \frac{1}{4}$,

giving $$\frac{1}{4} = \frac{3}{4}\left(x^2 - \frac{x^3}{3}\right)$$

$$\Rightarrow \quad x^3 - 3x^2 + 1 = 0.$$

This has approximate solution $x = 0.65$, and by symmetry the upper quartile will be at $x = 2 - 0.65 = 1.35$.

The inter-quartile range is then given by

$$1.35 - 0.65 \approx 0.70.$$

*Exercise 7D

1. Find the mode of these p.d.f.s:

(a) $f(x) = \frac{3}{50}\left(x^2 - 4x + 5\right) \quad 0 < x < 5$

(b) $f(x) = \frac{3}{13}\left(x^2 + 4\right) \quad 0 < x < 1.$

2. Find the median of the p.d.f.s given by:

(a) $f(x) = \frac{1}{8}(4 - x) \quad 0 < x < 4$

(b) $f(x) = e^{-x} \quad x > 0.$

7.6 Rectangular distribution

One very special continuous distribution that does not require calculus to analyse it is the rectangular or uniform distribution. Its p.d.f is defined as

$$f(x)=\begin{cases}\dfrac{1}{(b-a)} & \text{for } a<x<b\\ 0 & \text{otherwise}\end{cases}$$

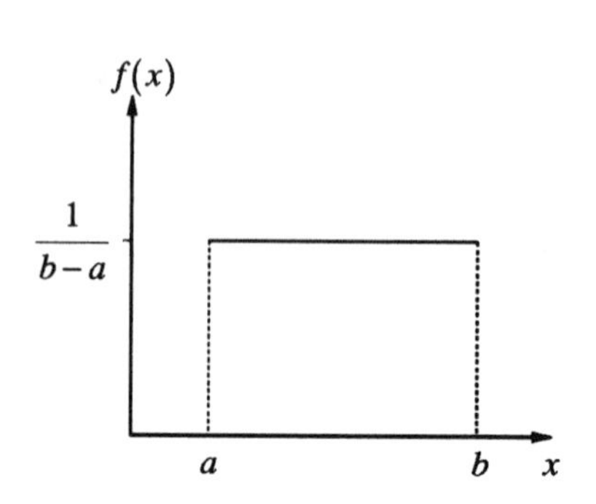

Its shape is illustrated opposite, and it is clear that its mean value is given by

$$E(X)=\frac{(a+b)}{2}$$

i.e. the midpoint of the line between a and b.

* Activity 9

(a) Use the formula to verify the formula for $E(X)$.

(b) By integration show that

$$V(X)=\frac{1}{12}(b-a)^2$$

Example

The continuous random variable X has p.d.f. $f(x)$ as shown opposite. Find

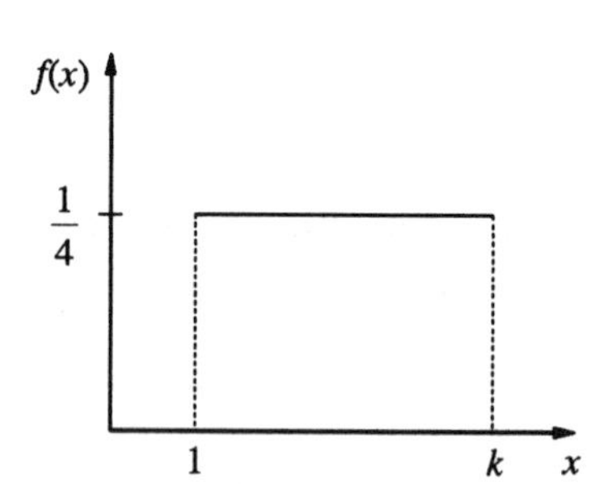

(a) the value of k

(b) $P(2.1<X<3.4)$

(c) $E(X)$

Solution

(a) The area under the curve must be 1, so

$$\frac{1}{4}\times(k-1)=1$$

$$\Rightarrow \quad k-1=4$$

$$\Rightarrow \quad k=5$$

(b) Now

$$P(2.1 < X < 3.4) = \text{area under the curve from } x = 2.1 \text{ to } x = 3.4$$

$$= \frac{1}{4} \times (3.4 - 2.1)$$

$$= \frac{1}{4} \times 1.3$$

$$= 0.325$$

(c) $E(X) = \frac{(1+5)}{2} = 3$, using the formula (or by symmetry)

Activity 10 How random are telephone numbers?

Take around a hundred telephone numbers as a group from a local telephone directory at random. Write down the last three digits in each number. These should be evenly spread in the range 000 to 999. (The first digits are often area codes.) Group them using group sizes of 200, i.e. 000-199 etc. and draw a histogram. Find the mean and variance of the data.

If the numbers were truly random and the sample sufficiently large, you would expect the distribution to be rectangular in shape. To form a p.d.f. you need to ensure that the total area of the graph is 1, so with a range of 1000 the p.d.f. is given by

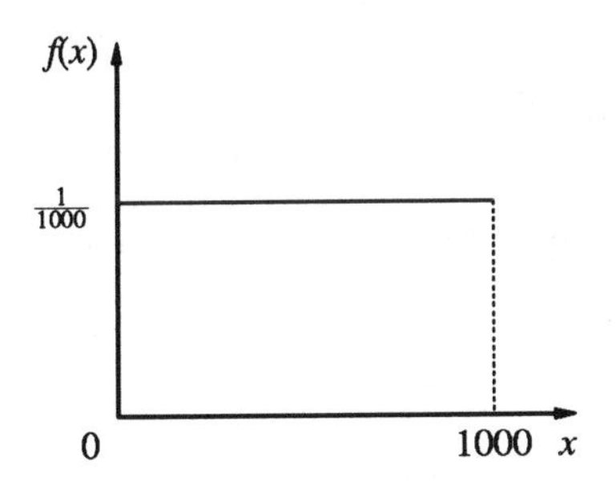

$$f(x) = \frac{1}{1000}, \quad 0 < x < 1000.$$

(In fact, for the data in Activity 10, only integer values are possible, but it is a reasonable approximation to use a continuous p.d.f.)

So the probability that a randomly chosen telephone number has the last three digits less than 300 is given by

$$\frac{300}{1000} = 0.3.$$

By symmetry the mean is given by $E(X) = 500$.

Using the formula, the variance is given by

$$V(x) = \frac{1}{12}(1000-0)^2$$

$$= \frac{1000000}{12}$$

$$= \frac{250000}{3}$$

$$\approx 8333$$

giving a standard deviation of

$$s \approx 91.30.$$

Check these against the values you obtained in your telephone survey.

*7.7 Miscellaneous Exercises

1. The distribution of petrol consumption at a garage is given by

$$f(x) = \begin{cases} ax^2(b-x) & 0 < x < 1 \\ 0 & \text{otherwise} \end{cases}$$

where x is in thousands of litres. Find the values of a and b if the mean consumption is 600 litres. Hence find the probability that in a given week the consumption exceeds 900 litres. (AEB)

2. A p.d.f is given by

$$f(x) = kx^2(3-x) \text{ for } 0 < x < 3.$$

Calculate

(a) the value of k to make this valid,

(b) the mean, μ, and variance, s^2, of the distribution and verify that $\mu + 2s = 3$,

(c) the probability that x differs from the mean by more than $2s$.

3. A continuous random variable, X, has p.d.f

$$f(x) = \begin{cases} x(x-1)(x-2) & 0 < x < 1 \\ a & 1 < x < 3 \\ 0 & \text{otherwise} \end{cases}$$

(a) Determine the value of a.

(b) Sketch the p.d.f.

(c) Find the value of $E(X)$ and $P(X < E(X))$.

What does this tell you about the median of the distribution?

4. A meat wholesaler sells remnants of meat in 5 kg bags. The amount in kg of inedibles (i.e. bone and gristle) is a random variable, X, with p.d.f.

$$f(x) = \begin{cases} k(x-1)(3-x) & 1 < x < 3 \\ 0 & \text{otherwise} \end{cases}$$

(a) Show that $k = \frac{3}{4}$.

(b) Find the mean and variance of X.

(c) Find the probability that X is greater than 2.5 kg. (AEB)

5. A small shopkeeper sells paraffin. She finds that during the winter the daily demand in gallons, X, may be regarded as a random variable with probability density function

$$f(x)=\begin{cases} kx^2(10-x) & 0\le x\le 10 \\ 0 & \text{otherwise} \end{cases}$$

(a) Verify that $k=0.0012$.

(b) Find the mean and the standard deviation of the distribution.

(c) Find the value of x which makes $f(x)$ a maximum. What is this value called?

(d) Estimate the median using the approximate relationship

$$2(\text{median}-\text{mean}) = \text{mode}-\text{median}.$$

Verify that your answer is approximately correct by finding the probability that an observation is less than your estimate of the median.

(e) If the shopkeeper has storage facilities for only eight gallons and can only replenish her stock once a day before the shop opens, find her mean daily sales. (AEB)

6. A teacher travels to work by car and the journey time, t hours, has a probability density function

$$f(t)=\begin{cases} 10ct^2 & 0\le t<0.6 \\ 9c(1-t) & 0.6\le t\le 1.0 \\ 0 & \text{otherwise} \end{cases}$$

where c is a constant.

(a) Find the value of c and sketch the graph of this distribution.

(b) Write down the most likely journey time taken by the teacher.

(c) Find the probability that the journey time will be

(i) more that 48 minutes;

(ii) between 24 and 48 minutes.

(AEB)

7. In a competition with a crossbow, contestants aim at a target with radius 5 cm. The target has a bull in the middle of it, of 2 cm radius. Hitting the bull scores 5 points and the outer circle 2. The p.d.f. of the variable X, the distance of a randomly fired shot from the centre of the target, is given by

$$f(x)=\begin{cases} 0.25e^{-0.25x} & x>0 \\ 0 & \text{otherwise} \end{cases}$$

Find

(a) the probability of hitting the bull,

(b) the probability of missing altogether, and

(c) hence the expected score of a single shot.

8. A p.d.f. is given by

$$f(x)=\frac{2}{3}\cos\left(x-\frac{\pi}{6}\right) \quad 0<x<\frac{2\pi}{3}.$$

(a) Show that this is a valid p.d.f.

(b) Find the mode of the distribution and hence sketch the curve.

(c) Find the probability that $x<1$.

9. The life of an electronic component is given by the p.d.f.

$$f(x)=\frac{100}{x^2} \text{ hours for } x>100.$$

Find

(a) the median life of a component.

(b) the probability that a component lasts for more than 250 hours.

8 THE NORMAL DISTRIBUTION

Objectives

After studying this chapter you should

- appreciate the wide variety of circumstances in which the normal distribution can be used;
- be able to use tables of the normal distribution to solve problems;
- be able to use the normal distribution as an approximation to other distributions in appropriate circumstances.

8.0 Introduction

The tallest accurately recorded human being was *Robert Wadlow* in the USA. On his death at the age of 22 he was 272 cm (8 feet 11.1 inches) tall. If you were an architect and you had to design doorways in a building you would clearly not make them all 9 feet high - most ceilings are lower than this!

What height should the ceilings be?

In 1980 the Government commissioned a survey, carried out on 10 000 adults in Great Britain. They found that the average height was 167.3 cm with SD (standard deviation) 9.1. You cannot make a door size that everyone can fit through but what height of door would 95% of people get through without stooping? This chapter should help you find the answer.

Activity 1 Data collection

There are many sets of data you could collect from people in your group, such as heights, weights, length of time breath can be held, etc. However, you will need about 100 results to do this activity properly so here are a few suggestions where large quantities of data can be collected quickly.

1. **Lengths of leaves**
   Evergreen bushes such as laurel are useful - though make sure all the leaves are from the same year's growth.

2. **Weights of crisp packets**
   Borrow a box of crisps from a canteen and weigh each packet accurately on a balance such as any Science laboratory would have.

3. **Pieces of string**
   Look at 10 cm on a ruler and then take a ball of string and try to cut 100 lengths of 10 cm by guessing. Measure the lengths of all the pieces in mm.

4. **Weights of apples**
   If anyone has apple trees in their garden they are bound to have large quantities in the autumn.

5. **Size of pebbles on a beach**
   Geographers often look at these to study the movement of beaches. Use a pair of calipers then measure on a ruler.

6. **Game of bowls**
   Make a line with a piece of rope on the grass about 20 metres away. Let everyone have several goes at trying to land a tennis ball on the line. Measure how far each ball is from the line.

Try at least two of these activities. You will need about 100 results in all. To look at the data it would help to have a data handling package on a computer.

8.1 Looking at your data

The data shown on the opposite page gives the length from top to tail (in millimetres) of a large group of frogs. This has been run through a computer package so you can see some useful facts about the data.

In the computer analysis you will see that most of the frogs are close to the mean value, with fewer at the extremes. This 'bell-shaped' pattern of distribution is typical of data which follows a normal distribution. To obtain a perfectly shaped and symmetrical distribution you would need to measure thousands of frogs.

Does your data follow a 'bell shaped' pattern?

You may notice that median $\approx$ mean $\approx$ mode, as might be expected for a symmetrical distribution. From the analysis of data you also see that the mean is 90.9 mm and the standard deviation is 11.7 mm. Now look at how much of the data is close to the mean, i.e. within one standard deviation of it. From the stem and leaf table you can see that 74 frogs have a length within one standard deviation above the mean and 59 within a SD below the mean.

Altogether, 133 frogs are +or – one SD from the mean, which is 66.5%.

Frog Data

The data below show the length from top to tail in millimetres of a large group of frogs.

83	69	97	53	89	95	105	80	76	117	74
91	100	77	110	68	118	87	97	78	100	95
73	103	96	72	71	99	121	81	104	68	89
87	96	87	72	79	102	98	97	88	87	86
103	79	104	105	91	82	102	75	95	90	62
65	97	86	97	111	98	92	74	88	84	80
95	96	92	95	100	90	91	95	75	70	84
80	98	96	94	101	85	113	96	103	98	95
84	84	97	95	108	94	79	81	92	85	87
90	85	82	81	97	79	90	90	94	98	73
91	91	107	102	89	85	98	84	91	90	86
113	86	93	77	100	96	90	97	109	102	84
85	87	97	92	107	102	104	94	93	75	96
91	117	91	87	118	96	89	88	111	120	92
76	94	104	80	77	94	84	78	73	92	81
83	104	91	91	96	88	115	96	74	88	86
80	98	101	95	96	102	78	97	80	87	82
72	78	108	91	91	91	110	86	101	81	97
82	97									

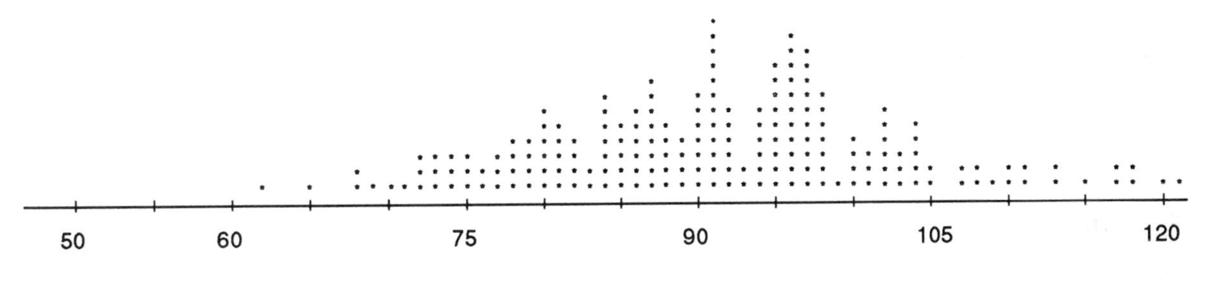

	N	MEAN	MEDIAN	TRMEAN	STDEV	MIN	MAX	Q1	Q3
Frogs	200	90.905	91.000	90.822	11.701	53.000	121.000	83.250	97.000

Stem and leaf of frogs

Leaf Unit = 1.0 N = 200

1	5	3
1	5	
2	6	2
6	6	5 8 8 9
17	7	0 1 2 2 2 3 3 3 4 4 4
33	7	5 5 5 6 6 7 7 7 8 8 8 8 9 9 9 9
57	8	0 0 0 0 0 0 1 1 1 1 1 2 2 2 2 3 3 4 4 4 4 4 4 4
85	8	5 5 5 5 5 6 6 6 6 6 6 7 7 7 7 7 7 7 7 8 8 8 8 8 9 9 9 9
(34)	9	0 0 0 0 0 0 0 1 1 1 1 1 1 1 1 1 1 1 1 2 2 2 2 2 2 3 3 4 4 4 4 4 4
81	9	5 5 5 5 5 5 5 5 5 6 6 6 6 6 6 6 6 6 6 6 7 7 7 7 7 7 7 7 7 7 7 7 8 8 8 8 8 8 8 9
41	10	0 0 0 0 1 1 1 2 2 2 2 2 2 3 3 3 4 4 4 4 4
20	10	5 5 7 7 8 8 9
13	11	0 0 1 1 3 3
7	11	5 7 7 8 8
2	12	0 1

Activity 2

Apply the same techniques to your own sets of data (i.e. draw up frequency tables or histograms and calculate means and SDs) and calculate the percentage which lie within one SD of the mean. If the data is **normally** distributed then this should be about 68%.

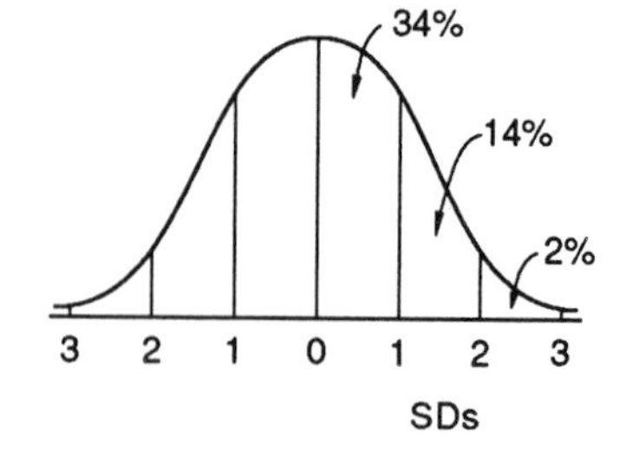

Similarly, you could look for the amount of data within 2 SDs, 3 SDs, etc. The table below gives approximately the percentages to expect.

Distances from mean in terms of standard deviation in one direction	0 − 1	1 − 2	2 − 3	over 3
Proportion of area in the above range	34 %	14 %	2 %	negligible

Note that very few items of data fall beyond three SDs from the mean.

What is clearly useful is that no matter what size the numbers are, if data are normally distributed, the proportions within so many SDs from the mean are always the same.

Example

IQ test scores, and the results of many other standard tests, are designed to be normally distributed with mean 100 and standard deviation 15.

Therefore statements such as the following can be made:

'68% of all people should achieve an IQ score between 85 and 115.'

'Only 2% of people should have an IQ score less than 70.'

'Only 1 in a 1000 people have an IQ greater than 145.'

Exercise 8A

The survey mentioned in the introduction also showed that the average height of 16-19 year olds was approximately 169 cm with SD 9 cm.

1. Assuming the data follows a normal distribution, find:

   (a) the percentage of sixth formers taller than 187 cm;

   (b) the percentage of sixth formers smaller than 160 cm;

   (c) in a sixth form of 300, the number of students smaller than 151 cm.

(Note these are not truly normal, as the pattern for girls and boys is different.)

8.2 The p.d.f. of the normal

If you could work in only whole numbers of SDs, the number of problems that could be solved would be limited. To calculate the proportions or probabilities of lying within so many SDs of the mean, you need to know the p.d.f. This was first discovered by the famous German mathematician, *Gauss* (1777-1855) and this is why the normal distribution is sometimes called the **Gaussian distribution.**
It is given by the formula

$$f(z) = \frac{1}{\sqrt{2\pi}} e^{-\frac{1}{2}z^2}.$$

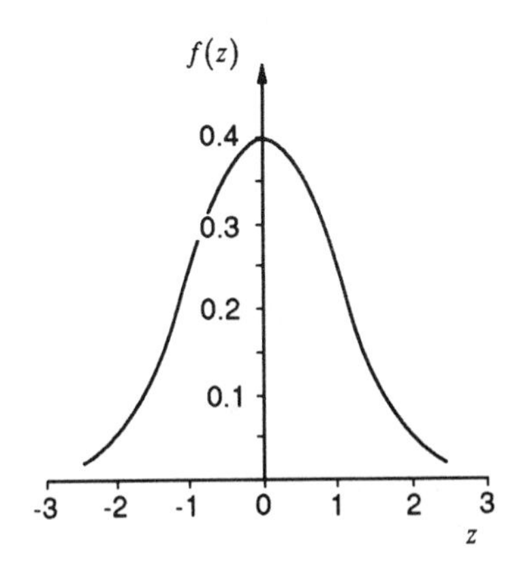

z is called the **standard normal variate** and represents a normal distribution with mean 0 and SD 1. The graph of the function is shown opposite.

Note that the function $f(z)$ has no value for which it is zero, i.e. it is possible, though very unlikely, to have very large or very small values occurring.

In order to find the probabilities of all possible SDs from the mean you would have to integrate the function between the values. This is a tedious task involving integration by parts and to avoid this tables of the function are commonly used. (See the table in the Appendix.)

z	.00	.01	.02	.03	.04	.05
0.0	.50000	.50399	.50798	.51197	.51595	.5199[illegible]
0.1	.53983	.54380	.54776	.55172	.55567	.559[illegible]
0.2	.57926	.58317	.58706	.59095	.59483	.598[illegible]
0.3	.61791	.62172	.62552	.62930	.63307	.636[illegible]
0.4	.65542	.65910	.66276	.66640	.67003	.673[illegible]
0.5	.69146	.69497	.69847	.70194	.70540	.708[illegible]
0.6	.72575	.72907	.73237	.73565	.73891	.74[illegible]
0.7	.75804	.76115	.76424	.76730	.77035	.7[illegible]
0.8	.78814	.79103	.79389	.79673	.79955	
0.9	.81594	.81859	.82121	.82381	.8[illegible]	
1.0	.84134	.84375	.84614	.84[illegible]		
1.1	.86433	.86650	.8686[illegible]			
1.2	.88493	.88686	.8[illegible]			
1.3	.90320	.90490	.[illegible]			
1.4	.91924	.92073				
1.5	.93319	.93448				
1.6	.94520	.94630				
1.7	.95543	.95637				
1.8	.96407	.964[illegible]				
1.9	.97128	.[illegible]				

$\Phi(1.0)$ (.84134); $\Phi(1.5)$ (.93319)

$$\Phi(z) = P(Z < z)$$
$$= \int_{-\infty}^{z} f(z)\, dz.$$

Here Φ has been used to denote the cumulative probability.

For positive z, the function gives you the probability of being less than z SDs above the mean.

For example, $\Phi(1.0) = 0.84313$, therefore 84.13% of the distribution is less than one SD above the mean.

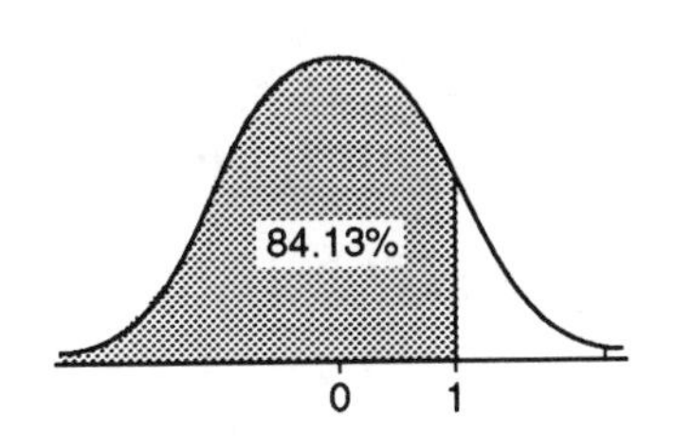

Tables usually give the area to the left of z and only for values above zero. This is because symmetry enables you to calculate all other values.

Example

What is the probability of being less than 1.5 SDs below the mean i.e. $\Phi(-1.5)$?

Solution

From tables,

$$\Phi(+1.5) = 0.9332$$

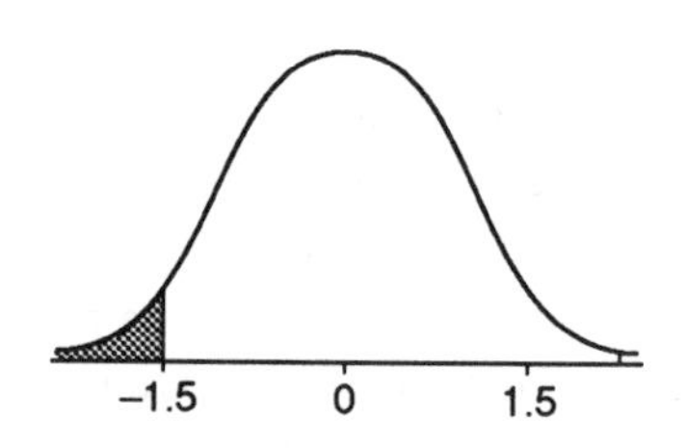

and by symmetry,

$$\Phi(-1.5) = 1 - 0.9332 = 0.0668$$

i.e. about 6.7%.

A random variable, X, which has this p.d.f. is denoted by

$$\boxed{X \sim N(0,1)}$$

showing that it is a normal distribution with mean 0 and standard deviation 1.

This is often referred to as the **standardised** normal distribution.

Example

If $X \sim N(0,1)$, find

(a) $P(x > 1.2)$

(b) $P(-2.0 < x < 2.0)$

(c) $P(-1.2 < x < 1.0)$

Solution

(a) $P(x > 1.2) = 1 - \Phi(1.2)$

$= 1 - 0.88493$ (from tables)

$= 0.11507$

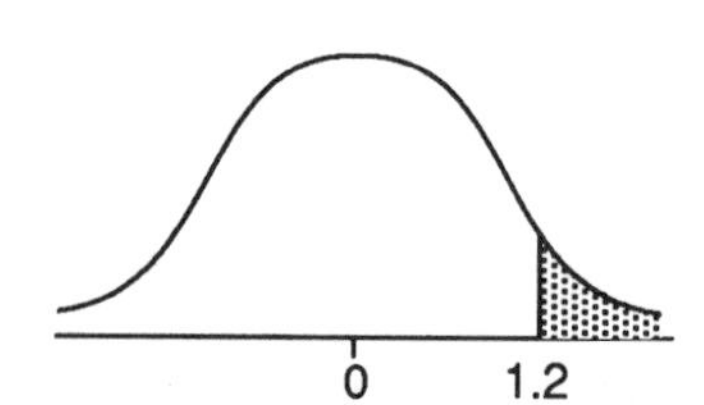

(b) $P(-2.0 < x < 2.0)$

$= P(x < 2.0) - P(x < -2.0)$

$= \Phi(2.0) - P(x > 2.0)$

$= \Phi(2.0) - (1 - P(x < 2.0))$

$= 2\Phi(2.0) - 1$

$= 2 \times 0.97725 - 1$

$= 0.9545$

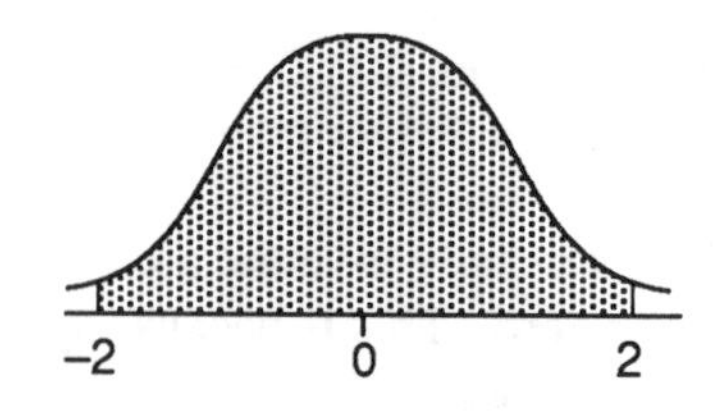

(c) $P(-1.2 < x < 1.0)$

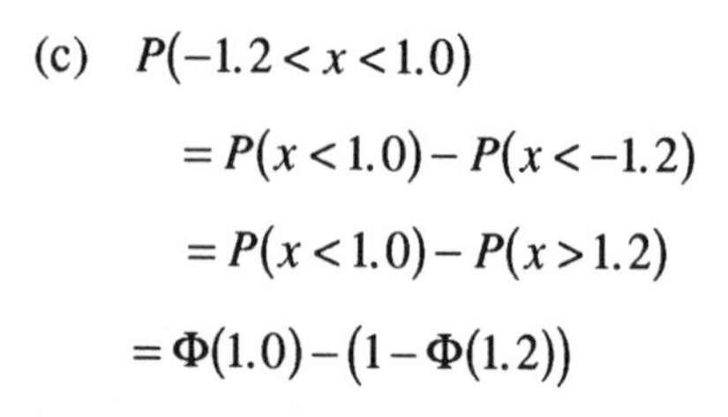

$= P(x < 1.0) - P(x < -1.2)$

$= P(x < 1.0) - P(x > 1.2)$

$= \Phi(1.0) - (1 - \Phi(1.2))$

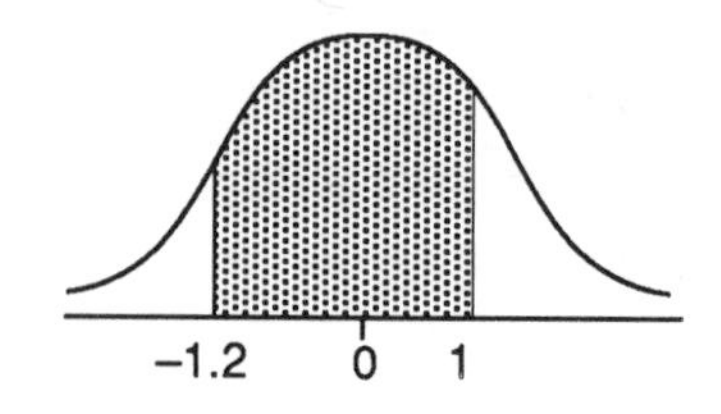

$= 0.84134 - (1 - 0.88493)$

$= 0.72627$

You can also use the tables to find the value of a when $P(X > a)$ is a given value and $X \sim N(0,1)$. This is illustrated in the next example.

Example

If $X \sim N(0,1)$, find a such that

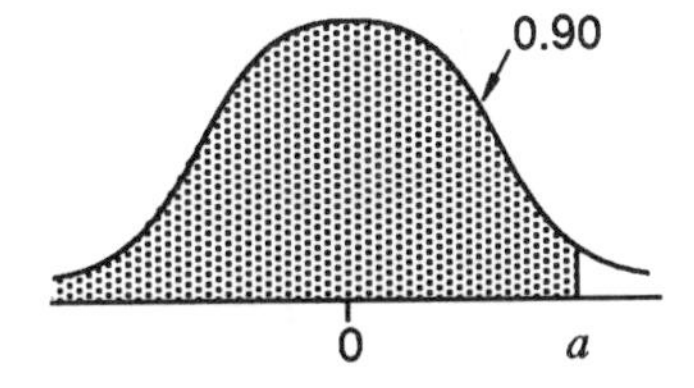

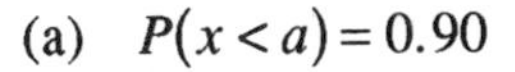

(a) $P(x < a) = 0.90$

(b) $P(x > a) = 0.25$

Solution

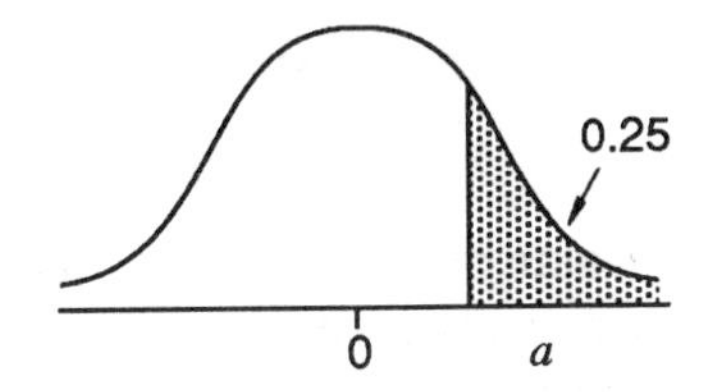

(a) Here $\Phi(a) = 0.90$, and from the tables

$$a \approx 1.28$$

(b) Here $\Phi(a) = 1 - 0.25 = 0.75$ and from the tables

$$a \approx 0.67$$

Exercise 8B

If $X \sim N(0,1)$, find

1. $P(x > 0.82)$
2. $P(x < 0.82)$
3. $P(x > -0.82)$
4. $P(x < -0.82)$
5. $P(-0.82 < x < 0.82)$
6. $P(-1 < x < 1)$
7. $P(-1 < x < 1.5)$
8. $P(0 < x < 2.5)$
9. $P(x < -1.96)$
10. $P(-1.96 < x < 1.96)$

8.3 Transformation of normal p.d.f.s

The method needed to transform any normal variable to the standardised variable is illustrated in the example below.

Example

Eggs laid by a particular chicken are known to have lengths normally distributed, with mean 6 cm and standard deviation 1.4 cm. What is the probability of:

(a) finding an egg bigger than 8 cm in length;

(b) finding an egg smaller than 5 cm in length?

Solution

(a) The number of SDs that 8 is above the mean is given by

$$z = \frac{x-\mu}{\sigma} = \frac{8-6}{1.4} = 1.429\,,$$

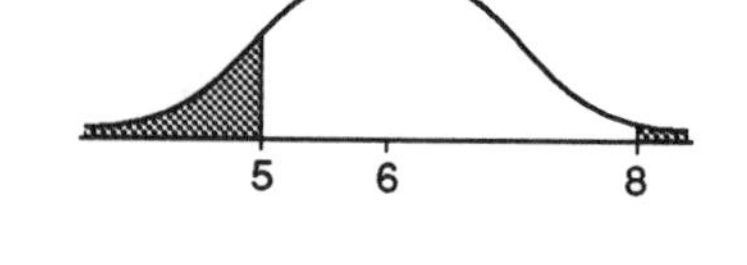

but $\Phi(1.43) = 0.9236$ (from tables)

so $P(x>8) = 1 - 0.9236 = 0.0764.$

(b) $$z = \frac{5-6}{1.4} = -0.7143,$$

but $\Phi(0.7143) = 0.7625$ (from tables using interpolation),

so $P(x<5) = 1 - 0.7625 = 0.2375.$

Note that using interpolation from tables is not necessary for the AEB examination, but it is good practice to use it to improve accuracy.

Note that in order to find the probability you need to establish whether you need the area greater than a half or less than a half. Drawing a diagram will help.

When a variable X follows a **normal distribution**, with mean μ and variance σ^2, this is denoted by

$$X \sim N\left(\mu, \sigma^2\right)$$

So in the last example, $X \sim N\left(6, 1.4^2\right)$.

To use normal tables, the transformation

$$z = \frac{x-\mu}{\sigma}$$

is used. This ensures that z has mean 0 and standard deviation 1, and the tables are then valid.

Using the UK data on heights in Section 8.0, the z value for Robert Wadlow's height is

$$z = \frac{272 - 167.3}{9.1} \approx 11.5.$$

So his height is 11.5 SDs above the mean. The most accurate tables show that 6 SDs is only exceeded with a probability of 10^{-10}, so it is extremely unlikely that a taller person will ever appear!

Example

If $X \sim N(4,9)$, find

(a) $P(x > 6)$

(b) $P(x > 1)$

Solution

Now $z = \dfrac{x-\mu}{\sigma} = \dfrac{x-4}{3}$,

(a) Hence

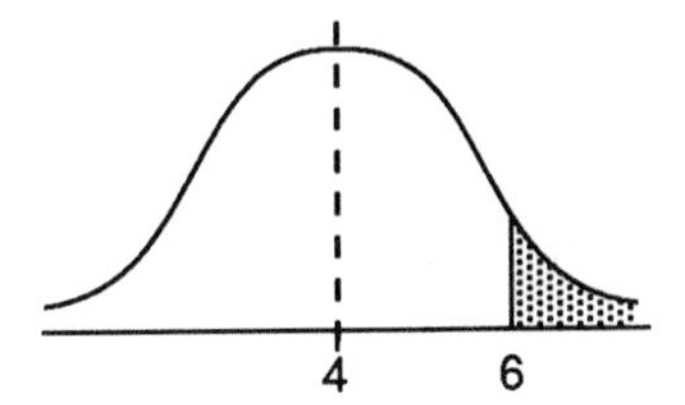

$$P(x > 6) = 1 - P(x < 6)$$
$$= 1 - \Phi\left(\frac{6-4}{3}\right)$$
$$= 1 - \Phi(0.67)$$
$$= 1 - 0.74857$$
$$= 0.25143$$

(b) $P(x > 1) = P(x < 7)$ (by symmetry)

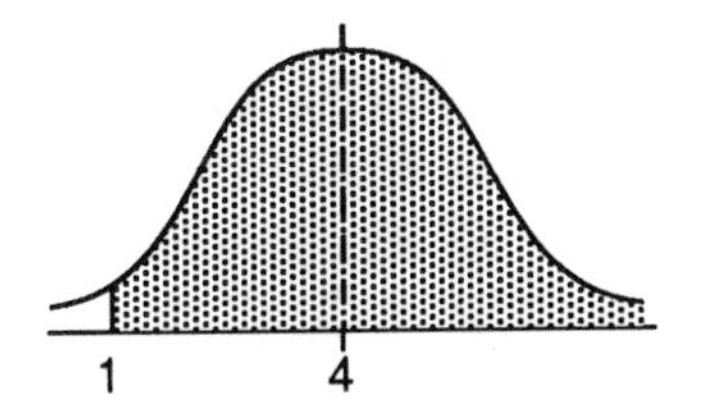

$$= \Phi\left(\frac{7-4}{3}\right)$$
$$= \Phi(1)$$
$$= 0.84134$$

Exercise 8C

1. If $X \sim N(200,625)$, find
   (a) $P(x > 250)$ (b) $P(-175 < x < 225)$
   (c) $P(x < 275)$

2. If $X \sim N(6,4)$, find
   (a) $P(x > 8)$ (b) $P(-8 < x < 8)$
   (c) $P(5 < x < 9)$

3. If $X \sim (-10,36)$, find
   (a) $P(x < 0)$ (b) $P(-12 < x < -8)$
   (c) $P(-15 < x < 0)$

4. Components in a personal stereo are normally distributed with a mean life of 2400 hours with SD 300 hours. It is estimated that the average user listens for about 1000 hours in one year. What is the probability that a component lasts for more than three years.

5. The maximum flow of a river in Africa during the 'rainy season' was recorded over a number of years and found to be distributed

   $N(6300,\ 1900^2)$ $m^3\ s^{-1}$.

   For the banks to burst a flow of 8700 $m^3\ s^{-1}$ is required. What is the probability of this happening in a particular year?

6. IQs are designed to be $N(100,\ 225)$. To join Mensa an IQ of 138 is required. What percentage of the population are eligible to join?
   A psychologist claims that any child with an IQ of 150+ is 'gifted'. How many 'gifted' children would you expect to find in a school of 1800 pupils?

7. Rainfall in a particular area has been found to be $N(850,\ 100^2)$ mm over the years. What is the probability of rainfall exceeding 1000 mm?

8. In a verbal reasoning test on different ethnic groups, one group was found to have scores distributed $N(98.42,\ 15.31^2)$. Those with a score less than 80 were deemed to be in need of help. What percentage of the overall group were in need of help?

8.4 More complicated examples

The following examples illustrate some of the many uses and applications of the normal distribution.

Example

A machine produces bolts which are $N(4,\ 0.09)$, where measurements are in mm. Bolts are measured accurately and any which are smaller than 3.5 mm or bigger than 4.4 mm are rejected. Out of a batch of 500 bolts how many would be acceptable?

Solution

$$P(X<4.4) = \Phi\left[\frac{(4.4-4)}{0.3}\right] \approx \Phi(1.33) = 0.9082$$

$$P(X<3.5) = \Phi\left[\frac{(3.5-4)}{0.3}\right] \approx \Phi(-1.67) = 0.0475.$$

Hence $P(3.5<X<4.4) \approx 0.9082 - 0.0475$

$$= 0.8607.$$

The number of acceptable items is therefore

$$0.8607 \times 500 = 430.35.$$

Example

IQ tests are measured on a scale which is $N(100,\ 225)$. A woman wants to form an 'Eggheads Society' which only admits people with the top 1% of IQ scores. What would she have to set as the cut-off point in the test to allow this to happen?

Solution

From tables you need to find z such that $\Phi(z) = 0.99$.

This is most easily carried out using a 'percentage points of the normal distribution' table (see Appendix), which gives the values directly.

Now $\Phi^{-1}(0.99) = 2.326$

which is an alternative way of saying that

$$\Phi(2.326) = 0.99.$$

(Check this using the usual tables.)

This means that

$$\frac{x - 100}{\sqrt{225}} = 2.326.$$

Hence $\quad x = 100 + (2.236 \times 15)$

$$= 133.54.$$

Example

A manufacturer does not know the mean and SD of the diameters of ball bearings he is producing. However, a sieving system rejects all bearings larger than 2.4 cm and those under 1.8 cm in diameter. Out of 1000 ball bearings 8% are rejected as too small and 5.5% as too big. What is the mean and standard deviation of the ball bearings produced?

Solution

Assume a normal distribution of

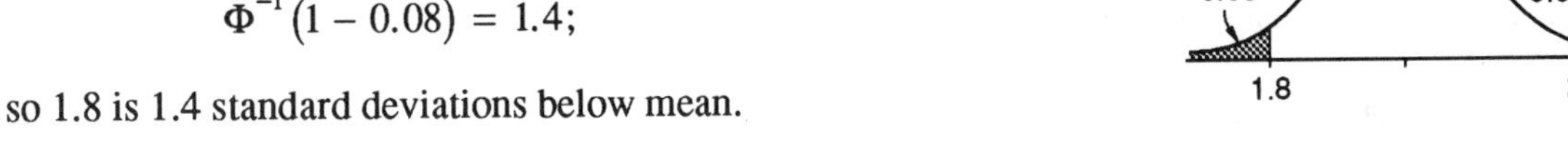

$$\Phi^{-1}(1 - 0.08) = 1.4;$$

so 1.8 is 1.4 standard deviations below mean.

Also $\quad \Phi^{-1}(1 - 0.055) = 1.6,$

so 2.4 is 1.6 standard deviations above the mean.

This can be written as two simultaneous equations and solved:

$$\mu + 1.6\sigma = 2.4$$

$$\mu - 1.4\sigma = 1.8.$$

Subtracting,

$$3.0\sigma = 0.6$$

$$\Rightarrow \quad \sigma = 0.2$$

Using the first equation,

$$\mu + (1.6 \times 0.2) = 2.4$$

$$\Rightarrow \quad \mu = 2.4 - (1.6 \times 0.2)$$

$$\Rightarrow \quad \mu = 2.08.$$

So diameters are distributed $N\left(2.08,\ 0.2^2\right)$.

Exercise 8D

1. Bags of sugar are sold as 1 kg. To ensure bags are not sold underweight the machine is set to put a mean weight of 1004 g in each bag. The manufacturer claims that the process works to a standard deviation of 2.4. What proportion of bags are underweight?

2. Parts for a machine are acceptable within the 'tolerance' limits of 20.5 to 20.6 mm. From previous tests it is known that the machine produces parts to $N(20.56, (0.02)^2)$.

   Out of a batch of 1000 parts how many would be expected to be rejected?

3. Buoyancy aids in watersports are tested by adding increasing weights until they sink. A club has two sets of buoyancy aids. One set is two years old, and should support weights according to $N(6.0, 0.64)$ kg; the other set is five years old and should support weights of $N(4.5, 1.0)$ kg. All the aids are tested and any which are unable to support at least 5 kg are thrown out.

   (a) If there are 24 two-year-old aids, how many are still usable?

   (b) If there are 32 five-year-old aids how many are still usable?

4. Sacks of potatoes are packed by an automatic loader with mean weight 114lb. In a test it was found that 10% of bags were over 116 lb. Use this to find the SD of the process. If the machine is now adjusted to a mean weight of 113 lb, what % are now over 116 lb if the SD remains unaltered?

5. In a soap making process it was found that $6\frac{2}{3}$ % of bars produced weighed less than 90.50 g and 4% weighed more than 100.25 g.

   (a) Find the mean and the SD of the process.

   (b) What % of the bars would you expect to weigh less than 88 g?

6. A light bulb manufacturer finds that 5% of his bulbs last more than 500 hours. An improvement in the process meant that the mean lifetime was increased by 50 hours . In a new test, 20% of bulbs now lasted longer than 500 hours.

   Find the mean and standard deviation of the original process.

8.5 Using the normal as an approximation to other distributions

In earlier chapters you looked at discrete distributions such as the binomial. Let us suppose that the probability of someone buying the *Daily Sin* newspaper in a particular town is 0.4. Consider these problems:

(a) What is the probability that in a row of six houses all six buy the *Sin*?

(b) Of 25 customers who come into a shop what is the probability of 10 or more buying the *Sin*?

(c) Two hundred people live on an estate. What is the probability that 100 or more buy the *Sin*?

In part (a) you would probably use the binomial distribution and a calculator to find $(0.4)^6$ and in (b) you would probably use tables to save on calculation. However, in part (c) there is a problem. Tables do not go beyond 50; you could use a Poisson approximation, but p is not really small and this would still involve enormous calculations. Imagine a probability histogram with 200 columns – it would look almost continuous! You will already know that for p approximately half, you get a symmetrical bell shaped graph. In fact you can use the normal distribution as an approximation in such cases.

You know that for a binomial distribution

$$\mu = np = 200 \times 0.4 = 80$$

and $$\sigma^2 = np(1-p) = 200 \times 0.4 \times 0.6 = 48$$

$$\Rightarrow \quad \sigma = 6.93.$$

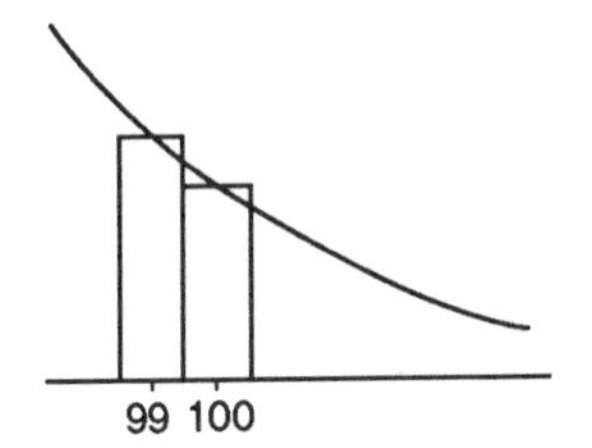

A slight adjustment needs to be made since the 100 column actually goes from 99.5 to 100.5. To include 100 you need to find $P(x > 99.5)$. This is sometimes called a **continuity correction** factor.

So P(100 or more buy *Sin*) $= 1 - \Phi\left(\dfrac{99.5 - 80}{6.93}\right)$

$$= 1 - \Phi(2.81)$$

$$= 1 - 0.99752 \quad \text{(from tables)}$$

$$= 0.00248.$$

In the same way you can use the normal distribution to approximate for the Poisson.

Example

Customers arrive at a garage at an average rate of 2 per five minute period. What is the probability that less than 15 arrive in a one hour period?

Solution

$\mu = \sigma^2 = 24$ per hour, so $\sigma = 4.9$.

Hence $P(\text{less than 15 in an hour}) = \Phi\left(\dfrac{14.5 - 24}{4.9}\right)$

$$\approx \Phi(-1.94)$$

$$= 1 - 0.97381$$

$$= 0.02619$$

(Note that 14.5 was used since **less** than 15 is required.)

Knowing when to use the normal distribution is important. Remember that it is only an approximation and if a simple calculation or tables will give the answer, this should be used.

You may have access to a computer package which can draw histograms of binomial and Poisson distributions for different n, p and λ, and overlay a normal distribution. The following diagrams show this for different cases.

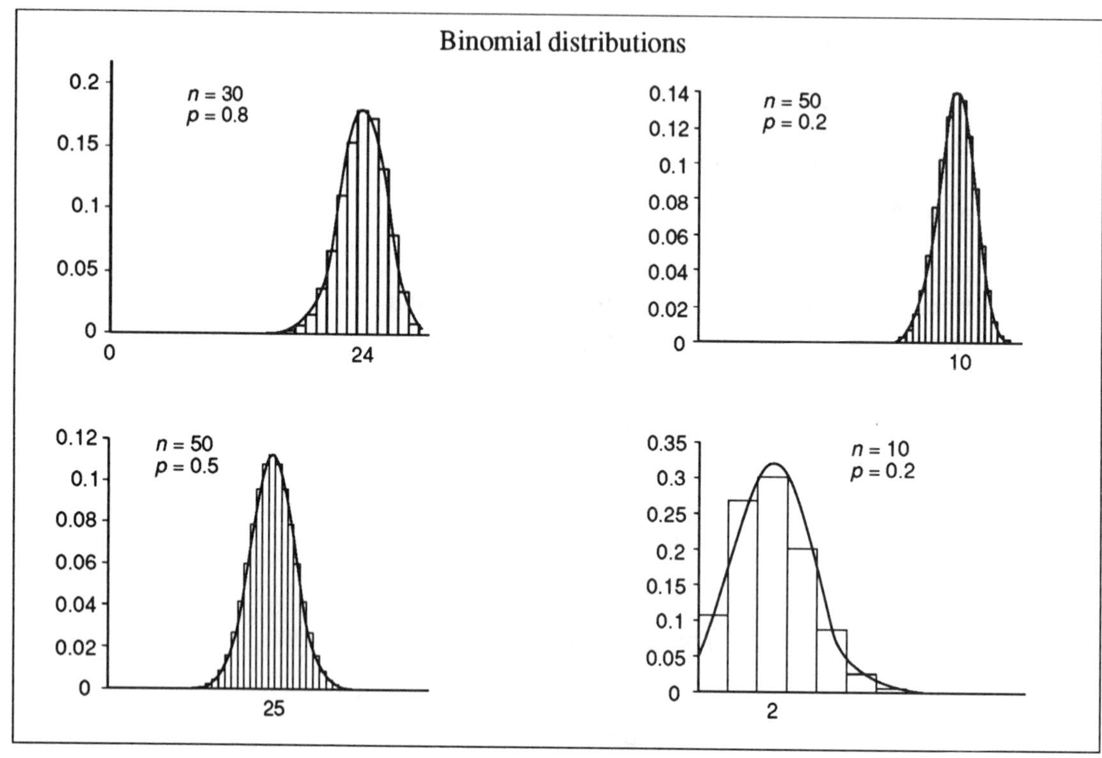

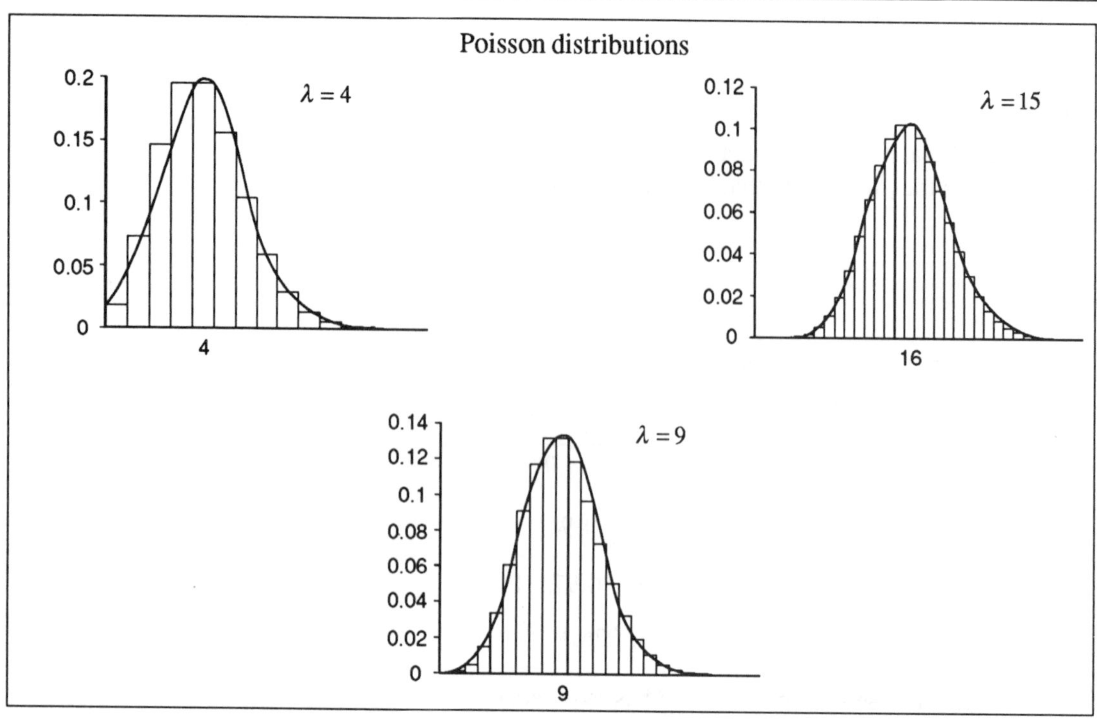

Activity 3

Check that the diagrams illustrate that

(a) for a binomial distribution, if p is close to 0.5, the normal is a good approximation even for quite small n. However, if p is small or large, then a larger value of n will be required for the approximation to be good;

(If $n > 30$, $np > 5$, $nq > 5$, then this is generally regarded as a satisfactory set of circumstances to use a normal approximation.)

(b) for a Poisson distribution, the larger n is the better the approximation.

($\lambda > 20$ is usually regarded as a necessary condition to use a normal approximation.)

To summarise, including the use of the Poisson to approximate to the binomial,

Distribution	Restrictions for using approximation	Approximating distribution
$X \sim B(n, p)$	n large (say >50) and p small (say <0.1)	$X \sim Po(np)$
$X \sim B(n, p)$	p close to $\frac{1}{5}$ and $n > 10$ or p moving away from $\frac{1}{2}$ and $n > 30$	$X \sim N(np, npq)$ $(q = 1 - p)$
$X \sim Po(\lambda)$	$\lambda > 20$ (say)	$X \sim N(\lambda, \lambda)$

Example

If $X \sim B(20, 0.4)$, find $P(6 \le X \le 10)$.

Also find approximations to this probability by using the

(a) normal distribution

(b) Poisson distribution.

Solution

$$P(X=6) = {}^{20}C_6(0.6)^{14}(0.4)^6 = 0.1244$$

Similarly $\quad P(X=7) = 0.1659$

$$P(X=8) = 0.1797$$

$$P(X=9) = 0.1597$$

$$P(X=10) = 0.1171$$

Hence $\quad P(6 \le X \le 10) = 0.747$ to 3 decimal places.

(a) Using a normal distribution,

$X \sim N(np, npq)$ where $np = 20 \times 0.4 = 8$

and $\quad npq = 20 \times 0.4 \times 0.6 = 4.8$

So

$$X \sim N(8, 4.8)$$

and $P(6 \le X \le 10) \rightarrow P(5.5 < X < 10.5)$.

With $z = \dfrac{x-8}{\sqrt{4.8}}$,

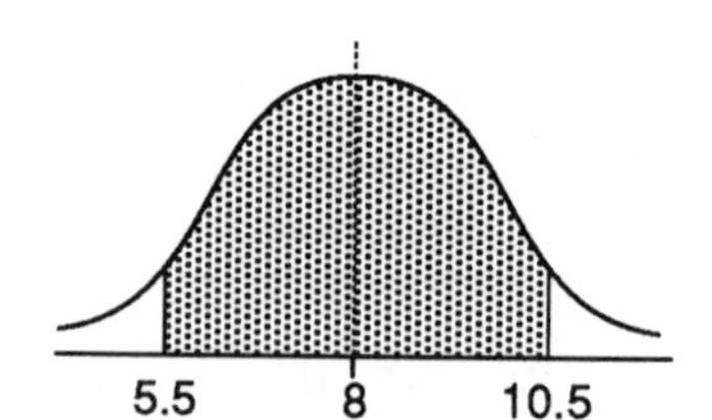

$$P(5.5 < X < 10.5) = \Phi\left(\frac{10.5-8}{\sqrt{4.8}}\right) - \Phi\left(\frac{5.5-8}{\sqrt{4.8}}\right)$$

$$= \Phi(1.141) - \Phi(-1.141)$$

$$= 2\Phi(1.141) - 1$$

$$\approx 2 \times 0.87286 - 1$$

$$= 0.746 \text{ to 3 decimal places.}$$

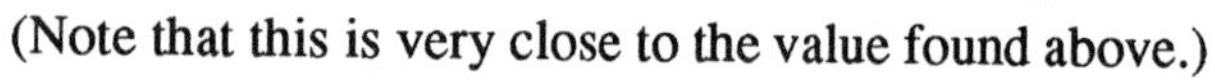

(Note that this is very close to the value found above.)

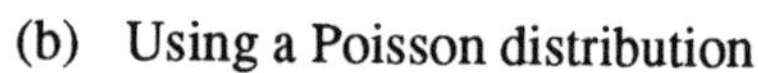

(b) Using a Poisson distribution

$$\lambda = np = 8$$

So

$$X \sim Po(8) \text{ and } P(X=x) = e^{-8}\frac{8^x}{x!}$$

This gives

$$P(X=6) = e^{-8}\frac{8^6}{6!} = 0.1221$$

Similarly $\quad P(X=7) = 0.1396$

$$P(X=8) = 0.1396$$

$P(X=9)=0.1241$

$P(X=10)=0.0993$

Thus $\quad P(6 \le X \le 10)=0.625$ to 3 decimal places.

(This is a poor approximation, since you should have $n>50$ and $p<\frac{1}{10}$ to use a Poisson approximation.)

Finally you should now be in a position to decide which of the distributions to use in order to model a situation.

Example

Answer the following questions using, in each case, tables of the binomial, Poisson or normal distribution according to which you think is most appropriate.

(a) Cars pass a point on a busy city centre road at an average rate of 7 per five second interval. What is the probability that in a particular five second interval the number of cars passing will be
 (i) 7 or less
 (ii) exactly 7?

(b) Weather records show that for a certain airport during the winter months an average of one day in 25 is foggy enough to prevent landings. What is the probability that in a period of seven winter days landings are prevented on
 (i) 2 or more days?
 (ii) no days?

(c) The working lives of a particular brand of electric light bulb are distributed with mean 1200 hours and standard deviation 200 hours. What is the probability of a bulb lasting more than 1150 hours?

(AEB)

Solution

(a) The Poisson distribution is suitable here since the question concerns a random event that can occur 0, 1, 2, ... times.

The mean value is $x=7$, giving, from tables,

(i) $P(7 \text{ or less})=0.5987$

(ii) $P(7)=P(7 \text{ or less})-P(6 \text{ or less})$

$=0.5987-0.4497$

$=0.149.$

(b) The binomial distribution is a suitable distribution with

$n = 7$ and $p = \dfrac{1}{25}$. Using tables,

(i) $P(2 \text{ or more}) = 1 - P(1 \text{ or less})$

$$= 1 - 0.9706$$
$$= 0.0294.$$

(ii) $P(\text{no days}) = \left(\dfrac{24}{25}\right)^7$

$$\approx 0.7514.$$

(c) The normal distribution is the model to use here, although the 'working lives' are not necessarily normal; so assume X, the working life, is distributed

$$X \sim N(1200,\ 200^2)$$

and

$$P(X > 1150) = 1 - P(X \le 1150)$$

$$= 1 - \Phi\left(\frac{1150 - 1200}{200}\right)$$
$$= 1 - \Phi(-0.25)$$
$$= \Phi(0.25)$$
$$= 0.59871.$$

Exercise 8E

1. The probability of someone smoking is about 0.4. What is the probability that:

   (a) in a group of 50 people more than half of them smoke;

   (b) in a group of 150, less than 50 of them smoke?

2. It is known nationally that support for the Story party is 32% from election results. In a survey carried out on 200 voters what is the probability that more than 80 of them are Story supporters?

3. A manufacturer knows from experience that his machines produce defects at a rate of 5%. In a day's production of 500 items 40 defects are produced. The Production Manager says this is not surprising. Is there evidence to support this?

4. Tickets for a concert are sold according to a Poisson distribution with mean 30 per day. What are the probabilities that:

   (a) less than 20 tickets are sold in one day;

   (b) all 180 tickets are sold in a five day working week?

5. Parts for a washing machine are known to have a weekly demand according to a Poisson distribution mean 20. How many parts should be stocked to ensure that a shop only runs out of parts on 1 in 20 weeks?

8.6 A very important application of the normal

Most modern calculators have a random number generator. The numbers produced generally follow a rectangular distribution in the range 0.000 to 0.999. These should therefore have mean 0.5, variance 0.083 $(\sigma = 0.289)$. (See Section 7.6)

Activity 4

Generate 10 random numbers and put them straight into the statistical function of your calculator. Write down x, the mean of your sample.

Repeat this 20 times and write down the means of the samples (remember to clear the statistical memories each time).

Plot these twenty results on normal probability paper and find the mean and SD of the sample means.

You should find that the twenty values are roughly normal, with mean, not suprisingly, 0.5 and SD 0.1. The SD has been decreased by a factor equivalent to the square root of the size of the sample, i.e. $\sqrt{10} = 3.16$.

This is the basis of a very important theorem, called the **Central Limit Theorem**. This says that, irrespective of the original distribution, sample means are normally distributed about the original distribution mean with 'standard error' equal to $\frac{\sigma}{\sqrt{n}}$, σ being the original SD and n the sample size. This will be explained in more detail in the next chapter.

8.7 Miscellaneous Exercises

1. The masses of plums from a certain orchard have mean 24g and standard deviation 5g. The plums are graded small, medium or large. All plums over 28g in mass are regarded as large and the rest equally divided between small and medium. Assuming a normal distribution find:

   (a) the proportion of plums graded large;

   (b) the upper limit of the masses of the plums in the small grade. (AEB)

2. A student is doing a project on the hire of videos from a local shop. She finds that the daily demand for videos is approximately normal, with mean 50 and SD 10.

   (a) What is the probability of more than 65 videos being hired on a particular day?

   (b) The shop is considering stopping the hire as it is uneconomical and decides that if demand is less than 40 on more than 3 days out of the next 7 it will do this.

   How likely is this to happen?

   (c) The student reckons that with a wider range of videos, demand would increase by 25% on average with no effect on the SD.

   What is the probability of more than 65 videos being hired if this happens?

3. A Dungeons & Dragons player is suspicious of a new die he has bought. He rolls the die 200 times and says he will throw it away if he gets more than 40 sixes. What is the probability of this happening with a fair die?

   A friend who is a Statistics student suggests that it would be better not to use 40 but to take a figure which a fair die would only exceed 5% of the time. What would this figure be?

4. In the survey of heights used earlier it was found that of males in the 16 - 19 year old age group 25% were taller than 178.8 cm and 10% were smaller than 165.4. Use this information to find the mean and SD of the distribution assuming it to be normal.

   What is the likelihood of a male in this age group being more than 183 cm (6 feet) tall?

5. Henri de Lade regularly travels from his home in the suburbs to his office in Paris. He always tries to catch the same train, the 08.05, from his local station. He walks to the station from his home in such a way that his arrival times form a normal distribution with mean 08.00 hours and SD 6 minutes.

   (a) Assuming that his train always leaves on time, what is the probability that, on any given day, Henri misses his train?

   (b) If Henri visits his office in this way 5 days each week and if his arrival times at the station each day are independent, what is the probability that he misses his train once, and only once, in a given week?

   (c) Henri visits his office 46 weeks every year. Assuming that there are no absences during this time, what is the probability that he misses his train less than 35 times in the year? (AEB)

6. The weights of pieces of home made fudge are normally distributed with mean 34 g and standard deviation 5 g.

   (a) What is the probability that a piece selected at random weighs more than 40g?

   (b) For some purposes it is necessary to grade the pieces as small, medium or large. It is decided to grade all pieces weighing over 40 g as large and to grade the heavier half of the remainder as medium. The rest will be graded as small. What is the upper limit of the small grade? (AEB)

7. Yuk Ping belongs to an athletics club. In javelin throwing competitions her throws are normally distributed with mean 41.0 m and standard deviation 2.0 m.

   (a) What is the probability of her throwing between 40 m and 46 m?

   (b) What distance will be exceeded by 60% of her throws?

   Gwen belongs to the same club. In competitions 85% of her javelin throws exceed 35 m and 70% exceed 37.5 m. Her throws are normally distributed.

   (c) Find the mean and standard deviation of Gwen's throws, each correct to two significant figures.

   (d) The club has to choose one of these two athletes to enter a major competition. In order to qualify for the final round it is necessary to achieve a throw of at least 48 m in the preliminary rounds. Which athlete should be chosen and why? (AEB)

8. Describe the main features of a normal distribution.

   A company has two machines cutting cylindrical corks for wine bottles. The diameters of corks produced by each machine are normally distributed. The specification requires corks with diameters between 2.91 cm and 3.12 cm.

   Corks cut on Machine A have diameters with a mean 3.03 cm and standard deviation 0.05 cm.

Calculate the percentage of corks cut on this machine that

(a) are rejected as undersize.

(b) meet the specification.

Machine B cuts corks with a mean diameter of 3.01 cm of which 1.7% are rejected as oversize. Calculate the standard deviation of the diameters of corks cut on Machine B.

Which machine, if either, do you consider to be the better? Explain. (AEB)

9. In parts (a) and (b) of this question use the binomial, Poisson or normal distribution according to which you think is the most appropriate. In each case draw attention to any feature of the data which supports or casts doubt on the suitability of the model you have chosen. Indicate, where appropriate, that you are using one distribution as an approximation to another.

(a) A technician looks after a large number of machines on a night shift. She has to make frequent minor adjustments. The necessity for these occurs at random at a constant average rate of 8 per hour. What is the probability that

(i) in a particular hour she will have to make 5 or fewer adjustments;

(ii)in an eight hour shift she will have to make 70 or more adjustments?

(b) A number of neighbouring allotment tenants bought a large quantity of courgette seeds which they shared between them. Overall 15% failed to germinate. What is the probability that a tenant who planted 20 seeds would have

(i) 5 or more failing to germinate;

(ii) at least 17 germinating? (AEB)

9 ESTIMATION

Objectives

After studying this chapter you should

- appreciate the importance of random sampling;
- understand the Central Limit Theorem;
- understand the concept of estimation from samples;
- be able to determine unbiased estimates of the variance;
- be able to find a confidence interval for the mean, μ.

9.0 Introduction

How will 'first time' voters cast their votes in a general election?

How do they differ from older voters?

Which issues concern them most?

Before these questions are considered it is worth noting a few ideas about statistics itself.

Firstly, if everyone was the same, there would be no need for statistics or statisticians; you could find out everything you needed to know from one person (or one event or one result). Statistics involves the study of variability so that estimates and predictions can be made in complex situations where there is no certain answer. The quality and usefulness of these predictions depend entirely on the quality of the data upon which they are based.

Activity 1

Consider again the three questions above.

Talk with other people in your group and decide:

(a) Which groups of people are referred to in the questions?

(b) How can each target group be defined? (i.e. How can you decide whether a person belongs to either group or not?)

(c) How can the information be obtained?

(d) Is it feasible to obtain information from all members of a population?

(e) Why might taking a sample/samples be a good idea?

(f) Could a sample survey possibly give better quality information than a census of the whole population?

9.1 Sampling methods

Methods of sampling have already been considered in Chapter 2; some of them will be revised again here. You will need the 'fish' sheet from Section 7.2.

Activity 2 Finding the mean by sampling

A : non random samples

(a) Select a **sample** of 5 fish which you think are representative.

(b) Measure the length of each fish in your sample (in mm).

(c) Calculate the mean length of the 5 fish in your sample and record your result.

(d) Repeat this for two more samples.

(e) Collect the results for everyone in your class and record them on a stem and leaf diagram (or frequency table).

B : random samples

Note that the fish are numbered from 1 to 57.

Use 2-figure random numbers from a random number table, calculator or computer to select a sample of 5 fish from the population. The method is described here. For 3-figure random numbers from a calculator, decide in advance whether you will use the first two digits, the last two digits, or the first and last.

Some of your 2-figure numbers will be larger than 57. These can be ignored without affecting the fairness of the selection process.

Here is an example showing a line of random numbers from a table:

25	82	33	06	74	18	34	09
	↓			↓			
	ignore			ignore			

The fish selected are numbered

25 33 6 18 and 34.

Note that you must use random numbers **consecutively** from the table after making a random start. You may **not** move about at will selecting numbers from different parts of the table.

Measure the lengths of these 5 fish as before.

Find the mean length of your sample.

Collect together sample means from all the students in your group. Display your results on a stem and leaf diagram.

Comparing sets of sample means

Compare the two stem and leaf diagrams for your sample means.

What do you notice?

Activity 3 Analysing the results

Answer the following questions with reference to your two sets of results from Activity 2. Firstly, though, measure the lengths of all fish on the sheet and find the true population mean, μ.

(a) How close were your results to the true population mean?

(b) How many samples under estimated μ ?

(c) How many samples over estimated μ ?

(d) Is either of your two sets of samples biased?

Definitions

To clarify your ideas, precise definitions will now be given.

Population

A **population** is the set of all elements of interest for a particular study. Quantities such as the population mean μ are known as **population parameters**.

Sample

A **sample** is a subset of the population selected to represent the whole population. Quantities such as the sample mean $\bar{x}$ are known as **sample statistics** and are **estimates** of the corresponding population parameters.

Random sample

A **random sample** is a sample in which each member of the population has an equal chance of being selected. Random samples generate **unbiased** estimates of the population mean, whereas non-random samples may not be unbiased. Also, the variability within random samples can be mathematically predicted (as the next section will show).

9.2 Sample size

The next experiment will consider the significance of the sample size. As the sample gets larger, so the estimate of the sample mean should become closer to the true population mean.

Activity 4 Selecting your samples

In this experiment you will need a table of random numbers or a calculator or computer to generate random numbers.

(a) Select a sample consisting of five single-digit random numbers (taking them consecutively from the random number table after making a random start). If you are using a computer or random number tables you require single-digit random numbers, so use each digit, one at a time. Treat three-figure random numbers from a calculator as three single numbers for your sample.

(b) Record these values together with their mean.

(c) Repeat this for four more samples (continuing to use consecutive random numbers).

(d) Now repeat the procedure for five samples each consisting of ten single-digit random numbers. Record your results together with their mean.

(e) Collect together the class results for means of samples of size $n = 5$ and $n = 10$ separately.

(f) Calculate the means of your two groups of sample means, and also the variances. Enter the values of $\bar{x}$ obtained for samples of size $n = 5$ into your calculator. Use the statistical functions to find $\bar{x}$, the mean of the sample means, and its standard deviation σ_n. Square this second result to find the variance of the $\bar{x}$s.

Repeat for samples of size $n = 10$.

(g) Which group of sample means is more variable?

Before any further analysis or discussion can be undertaken, the population mean and variance must be known.

The population of single-digit random numbers is theoretically infinite and consists of the numbers 0 to 9. These occur with equal probabilities and form a discrete uniform distribution, which you have already met in Chapter 4.

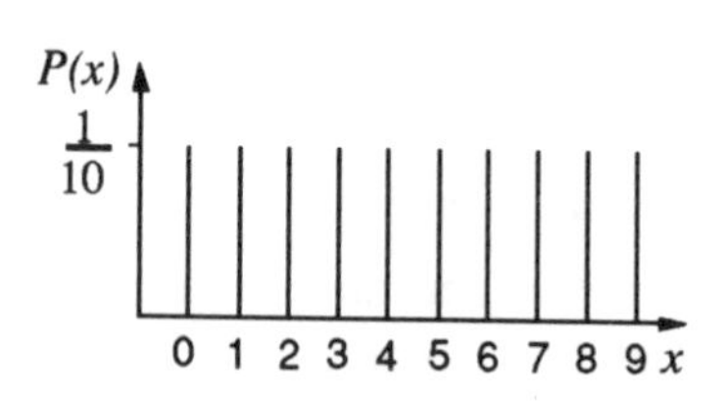

What is the value of $p(x)$ for $x = 0, 1, ..., 9$?

Activity 5 Exploring population parameters

(a) Use the formulae

$$\mu = \Sigma x p(x) \ , \quad \sigma^2 = \Sigma x^2 p(x) - \mu^2$$

to find the mean and variance of the population of single-digit random numbers.

(b) Do your class distributions for $\bar{x}(n = 5 \text{ and } n = 10)$ appear to be uniform distributions?

How would you describe them?

(c) Do any values of $\bar{x}$ appear to be more likely than others?

(d) Compare the mean of $\bar{x}$ with the population mean μ for $n = 5$ and $n = 10$.

(e) For samples of size $n = 5$, compare the variance of $\bar{x}$ with the population variance.

Is it close to $8.25 \div 5$?

(f) For samples of size $n = 10$, compare the variance of $\bar{x}$ with the population variance.

Is it close to $8.25 \div 10$?

9.3 The distribution of $\overline{X}$

Consider the idea of taking samples from a population. If it is a large population, it is possible (but perhaps not practical) to take a large number of samples, all of the same size from that population. For each sample, the mean $\bar{x}$ can be calculated. The value of $\bar{x}$ will vary from sample to sample and, as a result, is itself a random variable having its own distribution.

The value of the mean from any one sample is known as $\bar{x}$. If the distribution of all the possible values of the sample means is considered, this theoretical distribution is known as the **distribution of $\overline{X}$.**

So $\overline{X}$ itself is a random variable which takes different values for different random samples selected from a population.

In general, sample means are usually less variable though, than individual values. This is because, within a sample of size $n = 10$, say, large and small values in the sample tend to cancel each other out when $\bar{x}$ is calculated. In the example in the last section, even in a sample of ten random digits, $\bar{x}$ is unlikely to take a value greater than 7 or less than 2. Larger samples will generate values of $\overline{X}$ which are even more restricted in range (less variable).

There is an inverse relationship between the size of the samples and the variance of $\overline{X}$. Also, the distribution of $\overline{X}$ tends to be a peaked distribution (with mean and mode at μ) which approaches a normal distribution for large samples.

The Central Limit Theorem

The Central Limit Theorem describes the distribution of $\overline{X}$ if **all** possible random samples (of a given size) are selected from a population. The following results hold.

1. The mean of all possible sample means is μ the population mean; i.e.

$$E(\overline{X}) = \mu$$

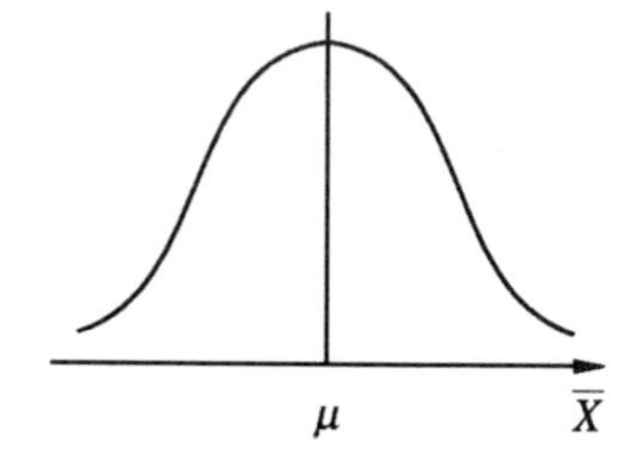

2. The variance of the sample means is the population variance divided by the sample size

$$V(\overline{X}) = \frac{\sigma^2}{n}$$

As n increases, the variance of $\overline{X}$ decreases and as $n \to \infty$

$$V(\overline{X}) \to 0.$$

3. If all possible values of $\overline{X}$ are calculated for a given sample size $n \geq 30$ a normal distribution is formed irrespective of the distribution of the original population: i.e. for $n \geq 30$

$$\overline{X} \sim N\left(\mu, \frac{\sigma^2}{n}\right)$$

Note that these results are true only for **random samples**. For non -random samples you cannot make predictions in terms of mean, variance or distribution of $\overline{X}$.

Use a computer package to investigate the distribution of $\overline{X}$ for random samples

(a) of different size, n;

(b) selected from different populations.

9.4 Identifying unusual samples

Afzal believes that the packets of crisps in the school tuck-shop are underweight. He takes a sample of ten packets of salt and vinegar crisps and finds their mean weight is 24.6 g. As the weight stated on the packets is 25 g, he writes to the manufacturer to complain. He receives the following reply :

> Dear Sir,
>
> Thank you for your letter of 5th July. We do share your concern over the weight of crisps in our packets of salt and vinegar crisps.
>
> Over a period of time, we have found that the standard deviation of the weights of individual packets is a little below 1 g. For this reason we believe that your sample mean weight of 24.6 g comes well within the normal limits of acceptability.
>
> Yours faithfully,

Does this reply give a valid argument?

The Central Limit Theorem can be used in practical situations like this to identify **unusual** samples, which are not typical of the population from which they have been selected.

For a given size of sample, the distribution of all possible sample means forms a normal distribution. The mean of this distribution is μ (the overall population mean) and the variance is $\dfrac{\sigma^2}{n}$

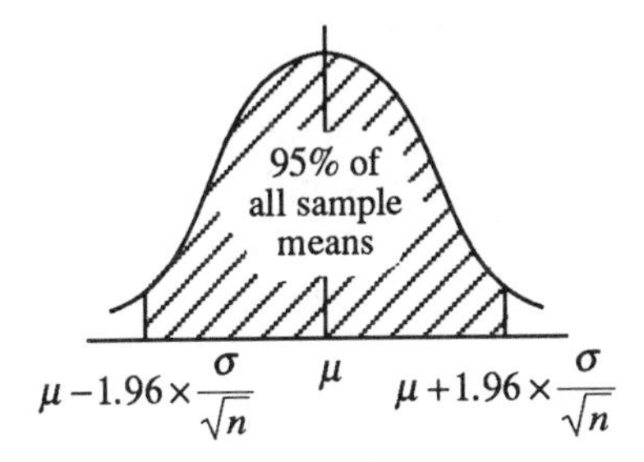

Distribution of all possible sample means for random samples of size n

Referring to normal distribution tables, 95% of any normal distribution lies between $z = -1.96$ and $z = +1.96$. So for a particular sample size n, 95% of all sample means should lie within 1.96 times the standard error each side of μ.

A sample mean outside this range may, in general, be :

(a) a genuine 'freak' result; after all, 5% of random samples do give means outside these limits;

(b) a result from a random sample selected from a different population;

(c) a sample selected from the population specified but not a random sample (e.g. a high proportion of children with above average IQs due to school selection procedures).

Example

A survey of adults aged 16-64 living in Great Britain, by the Office of Population Censuses and Surveys (OPCS), found that adult females had a mean height of 160.9 cm with standard deviation of 6 cm.

A sample of fifty female students is found to have a mean height of 162 cm. Are their heights typical of the general population?

Solution

The population mean is given by

$$\mu = 160.9 \text{ cm}.$$

Since the sample size is $n = 50$, the standard error is given by

$$\frac{\sigma}{\sqrt{n}} = \frac{6}{\sqrt{50}} = 0.849.$$

Thus the range of values for 95% of all sample means $(n = 50)$ is

$$160.9 - (1.96 \times 0.849) \leq \bar{x} \leq 160.9 + (1.96 \times 0.849)$$

$$160.9 - 1.66 \leq x \leq 160.9 + 1.66$$

So 95% of all $\bar{x}$ should lie in the range $159.24 \leq \bar{x} \leq 162.56$.

You can see that the sample mean of 162 cm obtained from the fifty students is within the range of typical values for $\bar{x}$. So there is no evidence to suggest that this sample is not typical of the population in terms of height.

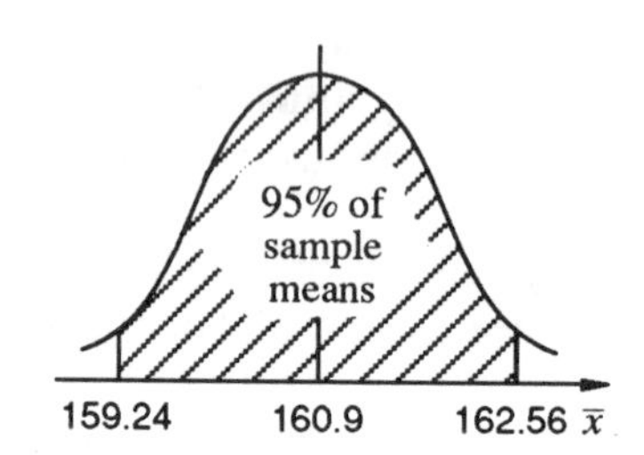

These ideas can be used to identify unusual sample means for large $(n \geq 30)$ random samples selected from any population, or for small samples selected from a normal population provided the value of the population variance is known.

Exercise 9A

1. IQ (Intelligence Quotient) scores are measured on a test which is constructed to give individual scores forming a normal distribution with a mean of 100 points and standard deviation of 15 points. A random sample of 10 students achieves a mean IQ score of 110 points. Is this sample typical of the general population?

2. A large group of female students is found to have a mean pulse rate (resting) of 75 beats per minute and standard deviation of 12 beats.

   Later, a class of 30 students is found to have a mean pulse rate of 82 beats per minute. What are your conclusions?

3. Over the summer months, samples of adult specimens of freshwater shrimps are taken from a slow moving stream. Their lengths are measured and found to have a mean of 39 mm and standard deviation of 5.3 mm. During the winter, a small sample of 10 shrimps is found to have a mean length of 41 mm.
   (a) Have the shrimps continued growing in the colder weather?
   (b) What assumptions have you had to make in order to answer the question?

4. Re-read the crisps problem at the beginning of this section. Do you agree with Afzal or do you agree with the manufacturers? Would it help Afzal to take a larger sample?

9.5 Confidence intervals

In many situations the value of μ, the population mean, may not be known for the variable being measured.

Is it possible to estimate the value of μ in such cases?

The best estimate of μ is the value of $\bar{x}$ obtained from a random sample. As the estimate consists of a single value, $\bar{x}$, it is referred to as a **point estimate**. (Other less reliable point estimates can be obtained from the sample median or mid-range.) The sample mean $\bar{x}$ is an **unbiased estimator** for μ, but even so, the value of $\bar{x}$ obtained from any particular random sample is unlikely to give the exact value of μ. In fact, as an unbiased estimator, half the values of $\bar{x}$ will under estimate μ, while half will give over estimates.

In order to 'hedge our bets' a range of values may be given which should include the value of μ. This is called an interval estimate or **confidence interval**.

The ideas introduced in earlier sections can be used to construct such an interval estimate.

Population variance known

The distribution of all possible sample means, $\bar{X}$, forms a normal distribution, with a mean μ, at the true population mean. (The variance of this distribution is $\frac{\sigma^2}{n}$ and decreases for larger sized samples.)

In reality, you are unlikely to know μ and all you have is one sample result $\bar{x}$. (Now $\bar{x}$ could lie anywhere in the distribution as shown in the diagram opposite.)

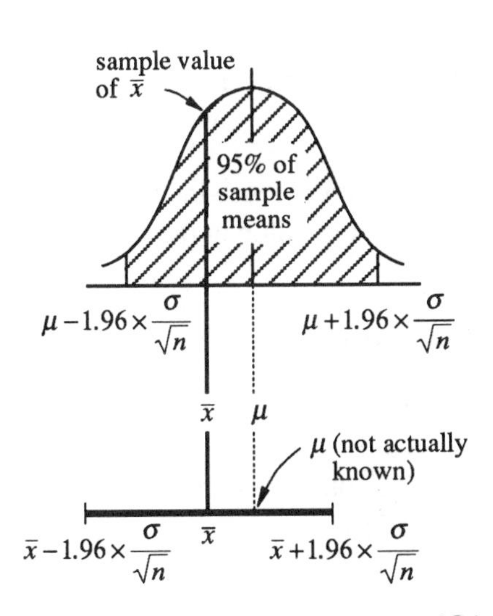

Distribution of all possible values of $\bar{x}$ obtained from random samples of size n.

In order to estimate μ, a range of values can be taken around $\bar{x}$ which hopefully will include the true value of μ.

The **95% confidence interval** for μ is found by taking a range of 1.96 times the standard error either side of $\bar{x}$; that is

$$\left(\bar{x} - 1.96\times\frac{\sigma}{\sqrt{n}} \quad , \quad \bar{x} + 1.96\times\frac{\sigma}{\sqrt{n}}\right).$$

Providing $\bar{x}$ lies within the central 95% of the distribution of all possible sample means, the confidence interval will include μ. This will happen for 95 random samples out of 100. If the sample is a 'freak' sample and the sample value of $\bar{x}$ lies at one of the extreme ends of the distribution, the confidence interval will not include μ. This will happen for only 5 random samples (roughly) out of every 100.

If this margin of error is to be reduced a wider interval (which will include μ for, say, 99 samples out of 100) can be constructed. This is called a **99% confidence interval.**

What values of the standard variable z trap 99% of the distributions?

Tables of the normal distribution give $z = \pm 2.58$ to give 0.5% of the distribution in each tail.

Since the standard error is $\frac{\sigma}{\sqrt{n}}$, the range of values for which 99% of all sample means should lie is

$$\left(\bar{x} - 2.58\times\frac{\sigma}{\sqrt{n}} \quad , \quad \bar{x} + 2.58\times\frac{\sigma}{\sqrt{n}}\right).$$

This gives the required confidence interval.

For 1 sample in every 100 the confidence interval will **not** include μ.

For other confidence intervals, e.g. 90%, 98%, you can look up the appropriate z value in the normal distribution tables.

Example

A random sample of 100 men is taken and their mean height is found to be 180 cm. The population variance $\sigma^2 = 49\ \text{cm}^2$. Find the 95% confidence interval for μ, the mean height of the population.

Solution

Lower limit $= \bar{x} - 1.96 \times \dfrac{\sigma}{\sqrt{n}}$

upper limit $= \bar{x} + 1.96 \times \dfrac{\sigma}{\sqrt{n}}$

when the standard error is given by

$$\frac{\sigma}{\sqrt{n}} = \sqrt{\frac{\sigma^2}{n}} = \sqrt{\frac{49}{100}} = 0.7$$

Hence the 95% confidence interval for μ is given by

$$= (180 \pm 1.96 \times 0.7) \text{ cm}$$

$$= (180 \pm 1.37) \text{ cm.}$$

So μ should lie between 178.63 and 181.37.

It should be noted at this stage that a **parameter** is a measure of a population, e.g. the population mean μ, or population variances, etc.; whilst a **statistic** is a similar measure taken from a sample, e.g. the sample mean $\bar{x}$.

So a statistic is an **estimator** of a parameter. When investigating a practical problem, it is unlikely that information concerning all the items in a given population will be available. Knowledge will normally be limited to one sample, from which tentative conclusions may be drawn concerning the whole population from which the sample is taken. The larger the sample, the greater the confidence in the estimation.

Activity 7

The lifetimes of 10 light bulbs were observed (in hours) as

1052 1271 836 962 1019 1051 512 1027 1219 1040

Assuming that the standard deviation for light bulbs of this type is 80 hours,

(a) find the 95% confidence interval for the mean lifetime of this type of bulb;

(b) find the % of the confidence interval that has a total range of 80 hours;

(c) determine the sample size, n, needed to restrict the range of the 95% confidence interval to 50 hours.

Population variance unknown

If the population variance is unknown and a large sample is taken, then the variance must be established from the sample itself. If s^2 is the **sample** variance, the best estimate for the **population** variance is given by

$$\hat{\sigma}^2 = \frac{ns^2}{(n-1)} \qquad \text{(this is derived in the text } \textit{Further Statistics}\text{)}$$

$$= \frac{n}{(n-1)}\left(\frac{1}{n}\Sigma x^2 - \bar{x}^2\right)$$

$$= \frac{1}{(n-1)}\left(\Sigma x^2 - n\bar{x}^2\right).$$

[This quantity is shown on calculators as σ_{n-1} or s_{n-1}.]

This result is used in the next example.

Example

A user of a certain gauge of steel wire suspects that its breaking strength, in newtons (N), is different from that specified by the manufacturer. Consequently the user tests the breaking strength, x N, of each of a random sample of nine lengths of wire and obtains the following *ordered* results.

72.2 72.9 73.4 73.8 74.1 74.5 74.8 75.3 75.9

$$\left[\Sigma x = 666.9 \quad \Sigma x^2 = 49\,428.25\right].$$

Calculate the mean and the variance of the sample values.

Hence calculate a 95% confidence interval for the mean breaking strength.

Comment upon the manufacturer's claims that the breaking strength of the wire has a mean of 75. (AEB)

Solution

For the sample,

$$\text{mean} = \frac{\Sigma x_i}{n}$$

$$= \frac{666.9}{9}$$

$$= 74.1$$

$$\text{variance} = \frac{1}{n}\Sigma x^2 - \bar{x}^2$$
$$= \frac{49428.25}{9} - (74.1)^2$$
$$\approx 1.218.$$

The estimate of the population variance is given by

$$\hat{\sigma}^2 = \frac{n}{(n-1)}s^2$$
$$= \frac{9}{8} \times 1.218$$
$$= 1.370.$$
$$\Rightarrow \quad \sigma \approx 1.170$$

The 95% confidence interval is now given by

$$\left(\bar{x} - 1.96 \times \frac{\hat{\sigma}}{\sqrt{n}} \quad , \quad \bar{x} + 1.96 \times \frac{\hat{\sigma}}{\sqrt{n}}\right)$$
$$\Rightarrow \quad \left(74.1 - 1.96 \times \frac{1.170}{\sqrt{9}} \quad , \quad 74.1 + 1.96 \times \frac{1.170}{\sqrt{9}}\right)$$
$$\Rightarrow \quad (73.34 \quad , \quad 74.86).$$

So the manufacturer's claim of a mean of 75 N is unlikely to be true since it is not included in the 95% confidence interval.

9.6 Miscellaneous Exercises

1. A sample of size 250 has mean 57.1 and standard deviation 11.8
   (a) Find the standard error of the mean.
   (b) Give 95% confidence limits for the mean of the population.

2. A company making cans for lemonade wishes to print 'Average contents x ml' on their cans, and to be 99% confident that the true mean volume is greater than x ml. The volume of lemonade in a can is known to have a standard deviation of 3.2 ml, and a random sample of 50 cans contained a mean volume of 503.6 ml.

   What volume of x should be stated?

3. Butter is sold in packs marked as salted or unsalted and the masses of the packs of both types of butter are known to be normally distributed. The mean mass of the salted packs of butter is 225.38 g and the standard deviation for both packs is 8.45 g.

   A sample of 12 of the unsalted packs of butter had masses, measured to the nearest gram, as follows.

   219 226 217 224 223 216 221 228 215 229 225 229

   Find a 95% confidence interval for the mean mass of unsalted packs of butter.

   Calculate limits between which 90% of the masses of salted packs of butter will lie.

   Estimate the size of sample which should be taken in order to be 95% sure that the sample mean of the masses of salted packs does not differ from the true mean by more than 3 g.

   State, giving a reason, whether or not you would use the same sample size to be 95% sure of the same accuracy when sampling unsalted packs of butter. (AEB)

4. The lengths of a sample of 100 rods produced by a machine are given below.

Length (cm)	5.60-5.62	5.62-5.64	5.64-5.66	5.66-5.68	5.68-5.70	5.70-5.72	5.72-5.74	5.74-5.76	5.76-5.78	5.78-5.80
Number of rods	1	3	5	5	8	20	24	16	12	6

Find the mean and standard deviation of the lengths in this sample.

Estimate the standard error of the mean, and give 95% confidence limits for the true mean length, μ, of rods produced by the machine. Explain carefully the meaning of these confidence limits.

By taking a larger sample, the manufacturers wish to find 95% confidence limits for μ which differ by less than 0.004 cm. Find the smallest sample size needed to do this.

5. A piece of apparatus used by a chemist to determine the weight of impurity in a chemical is known to give readings that are approximately normally distributed with a standard deviation of 3.2 mg per 100 g of chemical.

(a) In order to estimate the amount of impurity in a certain batch of the chemical, the chemist takes 12 samples, each of 100 g, from the batch and measures the weight of impurity in each sample. The results obtained in mg/100 g are as follows:

7.6 3.4 13.7 8.6 5.3 6.4

11.6 8.9 7.8 4.2 7.1 8.4

(i) Find 95% central confidence limits for the mean weight of impurity present in a 100 g unit from the batch.

(ii) The chemist calculated a 95% confidence interval for the mean weight of impurity in 100 g units from the batch. The interval was of the form $-\alpha \leq$ mean $\leq \alpha$.

Find the value of α.

Suggest why the chemist might prefer to use the value α rather than the limits in (i).

(iii) Calculate an interval within which approximately 90% of the measured weights of impurity of 100 g units from the batch will lie.

(b) Estimate how many samples of 100 g the scientist should take in order to be 95% confident that an estimate of the mean weight of impurity per 100 g is within 1.5 mg of the true value. (AEB)

6. Experimental components for use in aircraft engines were tested to destruction under extreme conditions. The survival times, X days, of ten components were as follows:

207 381 111 673 234 294 897 144 418 554

(a) Calculate the arithmetic mean and the standard deviation of the data.

(b) Assuming that the survival time, under these conditions, for all the experimental components is normally distributed with standard deviation 240 days, calculate a 90% confidence interval for the mean of X.

(AEB)

7. A company manufactures bars of soap. In a random sample of 70 bars, 18 were found to be mis-shaped. Calculate an approximate 99% confidence interval for the proportion of mis-shaped bars of soap.

Explain what you understand by a 99% confidence interval by considering

(a) intervals in general based on the above method,

(b) the interval you have calculated.

The bars of soap are either pink or white in colour and differently shaped according to colour. The masses of both types of soap are known to be normally distributed, the mean mass of the white bars being 176.2 g. The standard deviation for both bars is 6.46 g. A sample of 12 of the pink bars of soap had masses, measured to the nearest gram, as follows.

174 164 182 169 171 187
176 177 168 171 180 175

Find a 95% confidence interval for the mean mass of pink bars of soap.

Calculate also an interval within which approximately 90% of the masses of the white bars of soap will lie.

The cost of manufacturing a pink bar of soap of mass x g is $(15+0.065x)$p and it is sold for 32p. If the company manufactures 9000 bars of pink soap per week, derive a 95% confidence interval for its weekly expected profit from pink bars of soap. (AEB)

8. Sugar produced by a company is classified as granulated or caster and the masses of the bags of both types are known to be normally distributed. The mean of the masses of bags of granulated sugar is 1022.51 g and the standard deviation for both types of sugar is 8.21 g.

   Calculate an interval within which 90% of the masses of bags of granulated sugar will lie.

   A sample of 10 bags of caster sugar had masses, measured to the nearest gram, as follows.

   1062 1008 1027 1031 1011
   1007 1072 1036 1029 1041

   Find a 99% confidence interval for the mean mass of bags of caster sugar.

   To produce a bag of caster sugar of mass x g costs, in pence,

   $$(32 + 0.023x)$$

   and it is sold for 65p.

   If the company produces 10 000 bags of caster sugar per day, derive a 99% confidence interval for its daily profit from caster sugar.

   (AEB)

10 HYPOTHESIS TESTING

Objectives

After studying this chapter you should

- be able to define a null and alternative hypothesis;
- be able to calculate probabilities using an appropriate model to test a null hypothesis;
- be able to test for the mean based on a sample;
- understand when to use a one or two tailed test.

10.0 Introduction

One of the most important uses of statistics is to be able to make conclusions and test hypotheses. Your conclusions can never be absolutely sure, but you can quantify your measure of confidence in the result as you will see in this chapter.

Activity 1 Can you tell the difference?

Can you tell HP Baked Beans from a supermarket brand? Can you tell Coca Cola from a supermarket brand?

You are going to set up an experiment to determine whether people really can tell the difference between two similar foods or drinks.

Each person taking part in the test is given 3 samples: two of one product and one of another (so that they may have two cups containing Coca Cola (say) and one cup containing a supermarket brand or vice versa).

Ask the subject to identify the sample which is different from the other two.

Note that there are six possible groups of samples and a die can be used to decide which grouping to give to each individual subject taking part.

	(i)	(ii)	(iii)
1	A	B	B
2	B	A	B
3	B	B	A
4	B	A	A
5	A	B	A
6	A	A	B

Plan the experiment carefully before you start. Write out a list showing the samples and order of presentation for all your subjects (about 12, say).

Ensure that your subjects take the test individually in quiet surroundings, free from odours. All 3 samples must be of the same size and temperature. If there are any differences in colour you can blindfold your subject. Record each person's answer. Count the number of subjects giving the correct answer. Subjects who are unable to detect any difference at all in the 3 samples must be left out of the analysis.

10.1 Forming a hypothesis

In any experiment you usually have your own hypothesis as to how the results will turn out.

However it is usual to set up a **null hypothesis** that states the opposite of what you want to prove. This can only then be abandoned in the face of overwhelming evidence, thus placing the onus of proof on you.

The null hypothesis H_0

For the activity above your null hypothesis is that subjects cannot tell the difference between the 3 samples and that they are guessing.

The alternative hypothesis H_1

This is your experimental hypothesis (or what you really wish to prove). For the activity above your alternative hypothesis is that subjects really can distinguish between the samples (or some of them can at least).

These hypotheses can be written in mathematical terms as :

$$H_0 : p = \frac{1}{3}$$

$$H_1 : p > \frac{1}{3}$$

Here p is the probability of success assuming that H_0 is true; that is, subjects cannot tell the difference and are randomly guessing.

In order to reject H_0 and adopt H_1, your **experimental** results will have to be ones which are very difficult to explain under the null hypothesis.

You can use the binomial distribution to calculate the probabilities of people achieving various results by guesswork. Here are the probabilities for all the possible results for 10 subjects.

Number of people giving correct answer	Binomial probabilities
0	$\binom{10}{0}\left(\frac{2}{3}\right)^{10} = 0.0173$
1	$\binom{10}{1}\left(\frac{2}{3}\right)^{9}\left(\frac{1}{3}\right) = 0.0867$
2	$\binom{10}{2}\left(\frac{2}{3}\right)^{8}\left(\frac{1}{3}\right)^{2} = 0.1951$
3	$\binom{10}{3}\left(\frac{2}{3}\right)^{7}\left(\frac{1}{3}\right)^{3} = 0.2601$
4	$\binom{10}{4}\left(\frac{2}{3}\right)^{6}\left(\frac{1}{3}\right)^{4} = 0.2276$
5	$\binom{10}{5}\left(\frac{2}{3}\right)^{5}\left(\frac{1}{3}\right)^{5} = 0.1366$
6	$\binom{10}{6}\left(\frac{2}{3}\right)^{4}\left(\frac{1}{3}\right)^{6} = 0.0569$
7	$\binom{10}{7}\left(\frac{2}{3}\right)^{3}\left(\frac{1}{3}\right)^{7} = 0.0163$
8	$\binom{10}{8}\left(\frac{2}{3}\right)^{2}\left(\frac{1}{3}\right)^{8} = 0.0030$
9	$\binom{10}{9}\left(\frac{2}{3}\right)\left(\frac{1}{3}\right)^{9} = 0.0003$
10	$\binom{10}{10}\left(\frac{1}{3}\right)^{10} = 0.00002$

As the probability of 10 people guessing correctly is so small, if this actually happened you would be much more inclined to believe that they can actually tell the difference between the samples. So in this case it would be more rational to reject H_0, because the explanation offered by H_1 is more plausible.

How many of the other possible results are not easily explained by H_0 (and so better explained by H_1)?

Under H_0 the probability of :

10 correct guesses is	0.00002
9 or 10 correct guesses	0.00032
8, 9 or 10 correct guesses	0.00332
7, 8, 9 or 10 correct guesses	0.01962
6, 7, 8, 9 or 10 correct guesses	0.07652

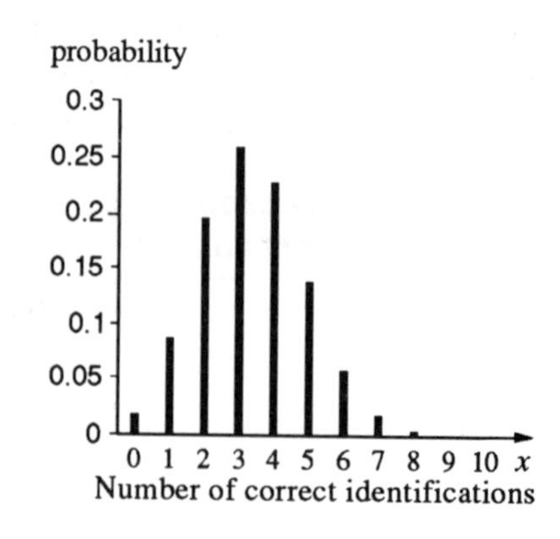

Note that if you adopt 9 correct as a 'significant' result you must include the probability for 10 as well (because 10 is actually a 'better' result than 9). Similarly with 8 you must include the probabilities for 9 or 10 correct and so on.

In scientific experiments, it is usual to take results with probabilities of 0.05 (5%) or less as convincing evidence for rejecting the null hypothesis.

If 10 subjects take the taste test then you will conclude that they *can* tell the difference between the samples if 7 or more of them make correct identifications.

Activity 2

Use a similar analysis to test your hypothesis in Activity 1.

Exercise 10A

1. A woman who claims to be able to tell margarine from butter correctly picks the 'odd' sample out of the 3 presented, for 5 out of 7 trials. Is this sufficient evidence to back up her claim?
2. A company has 40% women employees, yet of the 10 section heads, only 2 are women. Is this evidence of discrimination against women?
3. A subject takes a test for ESP (extra-sensory perception) in which he has to identify the suit of a playing card held by the experimenter. (The experimenter can see the card, but the subject cannot.) For 10 cards he makes 7 correct identifications. Is this evidence of ESP?

*10.2 The sign test

Another important use of hypothesis testing is to find out if, for a particular situation, you improve with practice.

Activity 3 Improve with practice?

(a) Maze

Time yourself finding your way through the maze shown opposite. Then have another try. Are you faster at the second attempt?

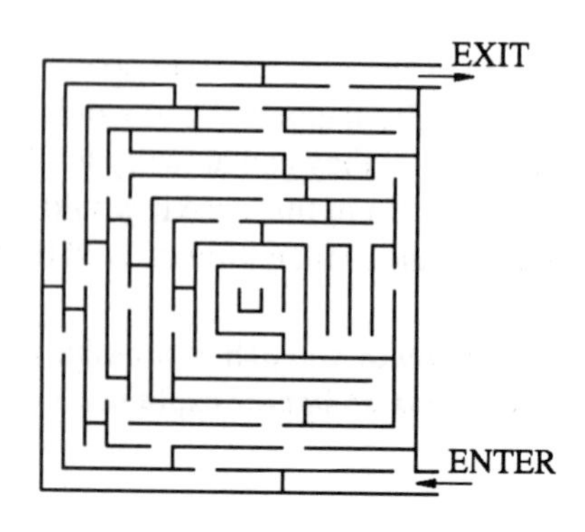

(b) Reaction times

Use a reaction ruler to find your reaction time. Record your first result and then your sixth (after a period of practice).

Analysing results

Suppose that 7 students took the maze test. Here are their times in seconds.

First try	Second try	Improvement?
9.0	3.5	✓
6.7	4.0	✓
5.8	2.6	✓
8.3	4.6	✓
5.1	5.4	X
4.9	3.7	✓
9.2	5.7	✓

Out of 7 subjects, 6 have improved, but could this result have occurred by chance? You can set up a hypothesis test in a similar way to the method used in the previous section.

Null hypothesis H_0

The null hypothesis is that any improvement or deterioration in times is quite random and that both are equally likely.

$$p(\text{improvement}) = \frac{1}{2}$$

or $$p(\text{deterioration}) = \frac{1}{2}$$

Alternative hypothesis H_1

In this situation you expect people to improve with practice so the alternative hypothesis is that

$$p(\text{improvement}) > \frac{1}{2}$$

In order to analyse the experimental results ignore any students who manage to achieve identical times in both trials. Their results actually do not affect your belief in the null hypothesis either way. If there are n students with non-zero differences in times, the binomial distribution can be used as a model to generate probabilities under the null hypothesis. If x is the random variable 'number of positive differences', then $X \sim B(7, \frac{1}{2})$.

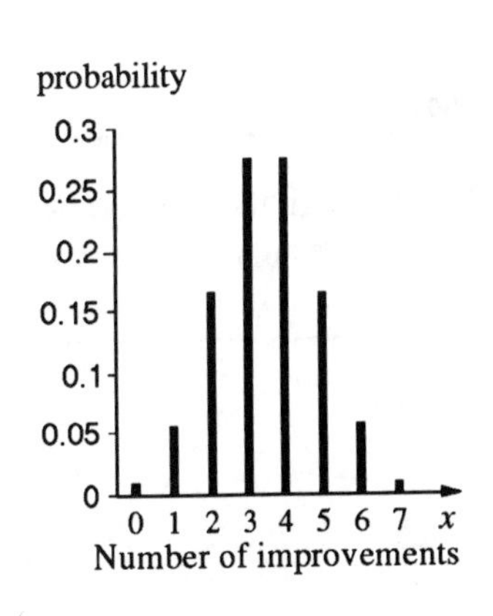

This gives the table of probabilities shown below

Number of positive differences	Probability	
0	$\binom{7}{0}\left(\frac{1}{2}\right)^7$	$= 0.0078$
1	$\binom{7}{1}\left(\frac{1}{2}\right)^7$	$= 0.0547$
2	$\binom{7}{2}\left(\frac{1}{2}\right)^7$	$= 0.1641$
3	$\binom{7}{3}\left(\frac{1}{2}\right)^7$	$= 0.2734$
4	$\binom{7}{4}\left(\frac{1}{2}\right)^7$	$= 0.2734$
5	$\binom{7}{5}\left(\frac{1}{2}\right)^7$	$= 0.1641$
6	$\binom{7}{6}\left(\frac{1}{2}\right)^7$	$= 0.0547$
7	$\binom{7}{7}\left(\frac{1}{2}\right)^7$	$= 0.0078$

The binomial probabilities have been calculated according to $H_0 : X \sim B(7, \frac{1}{2})$. Under H_0, the probability of :

7 improvements is	0.0078
6 or 7 improvements is	0.0625

If you adhere to a 5% level of significance (0.05 probability of rejecting H_0 when it may be true), the result of 6 improvements is actually not sufficient grounds for rejecting H_0. This method of hypothesis testing is called the **sign test**.

Activity 4

Follow through the method outlined in the previous section to analyse the results of your experiments in Activity 3.

*Exercise 10B

In all questions, assume a 5% level of significance.

1. A group of students undertook an intensive six week training programme, with a view to improving their times for swimming 25 metres breast-stroke. Here are their times measured before and after the training programme. Have they improved significantly?

25m breast stroke times in seconds

student	A	B	C	D	E	F	G	H
before programme	26.7	22.7	18.4	27.3	19.8	20.2	25.2	29.8
after programme	22.5	20.1	18.9	24.8	19.5	20.9	24.0	24.0

2. Twelve young children (6-year-olds) were given a simple jigsaw puzzle to complete. The times they took were measured on their first and second attempts. Did they improve significantly?

jigsaw puzzle times in seconds

child	1	2	3	4	5	6
first attempt	143	43	271	63	232	51
second attempt	58	45	190	49	178	58

child	7	8	9	10	11	12
first attempt	109	156	304	198	83	115
second attempt	73	127	351	170	74	97

3. A group of 9 children wanted to see whether the amount of air in their bicycle tyres made a difference in how easy it was to pedal their bikes. They decided to ride a particular route under two different conditions : once with a tyre pressure of 40 pounds per square inch (psi) and once with 65 psi. (The order in which they did this was to be decided by tossing a coin.) The time it took (in minutes) for each circuit was :

40 psi	34	54	23	67	46	35	49	51	27
65 psi	32	45	21	63	37	40	51	39	23

Are the children significantly faster with the higher pressure tyres?

4. A group of engineering students run a test to see whether cars will get as many mpg on lead-free petrol as on 4-star petrol.

car	A	B	C	D	E	F	G	H	I	J
lead-free	15	23	21	35	42	28	19	32	31	24
4-star	18	21	25	34	47	30	19	27	34	20

Does 4-star petrol give significantly better results?

5. A personnel director of a large company would like to know whether it will take less time to type a standard monthly report on a word processor or on a standard electric typewriter. A random sample of 7 secretaries was selected and the amount of typing time recorded in hours.

secretary	A	B	C	D	E	F	G
word processor	6.3	7.5	6.8	6.0	5.3	7.4	7.2
electric typewriter	7.0	7.4	7.8	6.7	6.1	8.1	7.5

Are the secretaries significantly faster using the word processors?

10.3 Hypothesis testing for a mean

You can now extend the ideas, introduced in earlier sections, to the testing of a hypothesis about the mean of a sample. There are two cases to consider, firstly tests for the mean based on a sample from a normal distribution with known variance, and secondly tests based on a large sample from an unspecified distribution.

Example

Afzal weighs the contents of 50 more packets of crisps and finds that the mean weight of his sample is 24.7 g. The weight stated on the packet is 25 g and the manufacturers claim that the weights are normally distributed with standard deviation 1 g. Can Afzal justifiably complain that these packets are underweight?

Solution

For this problem

$$H_0 : M = 25\text{g}$$

$$H_1 : M < 25\text{g}$$

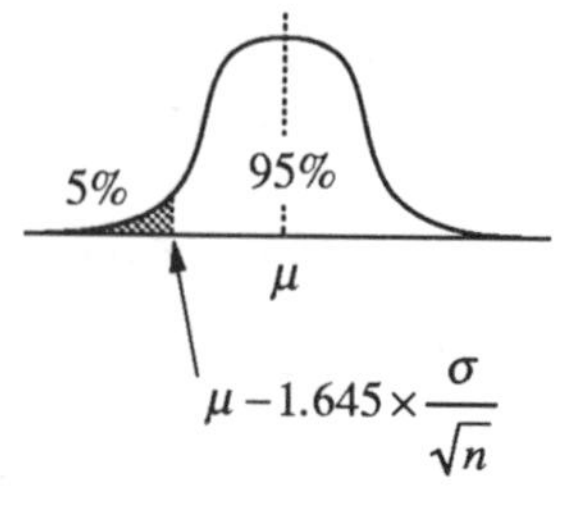

As Afzal suspects that the crisps are underweight he will reject the null hypothesis for unusually low values of $\bar{x}$. The critical region consists of these values at the extreme left hand end of the distribution of $\bar{x}$, which have a 5% probability in total. (This is called a **one tailed test**.)

The critical value of z, which can be found from normal distribution tables, is -1.645.

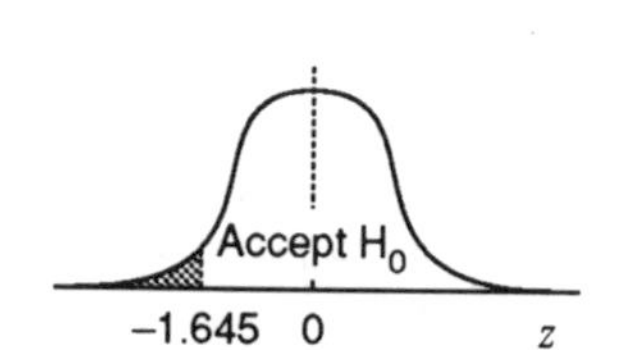

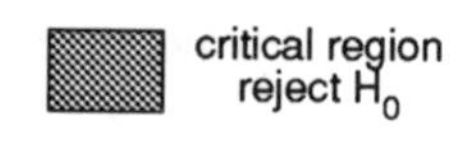

Under H_0, $\quad \bar{X} \sim N\left(25, \frac{1}{50}\right)$

Now the test statistic is

$$z = \frac{\bar{x} - \mu}{\text{standard error}}$$

$$= \frac{\bar{x} - \mu}{\left(\frac{\sigma}{\sqrt{n}}\right)}$$

$$= \frac{24.7 - 25}{\left(\frac{1}{\sqrt{50}}\right)}$$

$$= -2.12$$

This value of z is significant as it is less than the critical value, -1.645, and falls in the critical region for unusual values of $\bar{x}$. As it is extremely unlikely under H_0 (and is better explained by H_1) you can reject H_0. Afzal's results are such that he has good cause to complain to the manufacturers!

Example 2

A school dentist regularly inspects the teeth of children in their last year at primary school. She keeps records of the number of decayed teeth for these 11-year-old children in her area. Over a number of years, she has found that the number of decayed teeth was approximately normally distributed with mean 3.4 and standard deviation 2.1.

standard deviation 2.1.

She visits just one middle school in her rounds. The class of 28 12-year-olds at that school have a mean of 3.0 decayed teeth. Is there any significant difference between this group and her usual 11-year-old patients?

Solution

For this problem

$$H_0 : M = 3.4$$

$$H_1 : M \neq 3.4$$

The dentist has no reason to suspect either a higher or lower figure for the mean for 12-year-olds. (Children at this age may still be losing milk teeth) so the alternative hypothesis is non directional and a **two tailed test** is used.

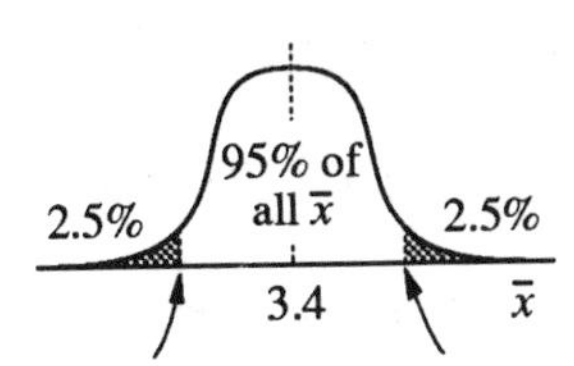

The critical region (consisting of unusual results with low probabilities) is split, with $2\frac{1}{2}\%$ at both extremes of the distribution. The critical values of z are ± 1.96, from normal distribution tables.

As before, the test statistic is

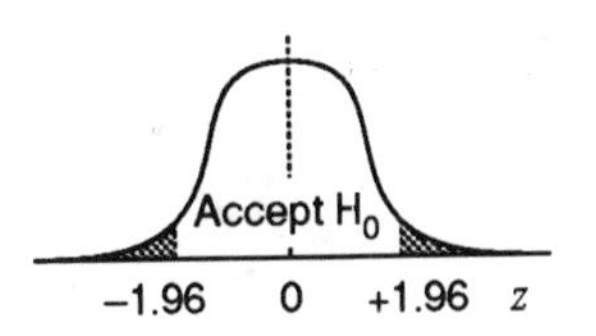

$$z = \frac{\bar{x} - \mu}{\left(\frac{\sigma}{\sqrt{n}}\right)}$$

$$= \frac{3.0 - 3.4}{\left(\frac{2.1}{\sqrt{28}}\right)}$$

$$= -1$$

This value of z is not significant. It lies well within the main body of the distribution for $\bar{x}$. You must accept H_0 and conclude that the result for the 12-year-olds is not unusual.

When the distribution is unknown and the variance, σ^2, unknown, you have to use the Central Limit Theorem which states that the distribution of the mean of sample is normally distributed,

$$\bar{X} \sim N\left(\mu, \frac{\sigma^2}{n}\right).$$

Since σ^2 is unknown, the estimate

$$\hat{\sigma}^2 = \frac{ns^2}{n-1}$$

is used, when s^2 is the sample variance. But for larger n,

$$\frac{n}{n-1} \approx 1$$

and so $\hat{\sigma}^2 = s^2$, and you use the test statistics

$$z = \frac{\bar{x} - \mu}{\left(\frac{s}{\sqrt{n}}\right)}.$$

Example

A manufacturer claims that the average life of their electric light bulbs is 2000 hours. A random sample of 64 bulbs is tested and the life, x, in hours recorded. The results obtained are as follows:

$$\Sigma x = 127\,808 \qquad \Sigma(\bar{x} - x)^2 = 9694.6$$

Is there sufficient evidence, at the 1% level, that the manufacturer is over estimating the length of the life of the light bulbs?

Solution

From the sample

$$\bar{x} = \frac{\Sigma x}{n} = \frac{127\,808}{64} = 1997$$

$$s^2 = \frac{\Sigma(\bar{x} - x)^2}{n} = \frac{9694.6}{64} = 151.48$$

giving the sample standard deviation as

$$s = 12.31.$$

Let X be the random variable, the life (in hours) of a light bulb, so define

$$H_0 : \mu = 2000$$

$$H_1 : \mu < 2000 \quad \text{(assuming the manufacturer is over estimating the lifetime)}$$

Assuming H_0,

$$\bar{X} \sim N\left(\mu, \frac{\sigma^2}{n}\right) \approx N\left(2000, \frac{151.48}{64}\right).$$

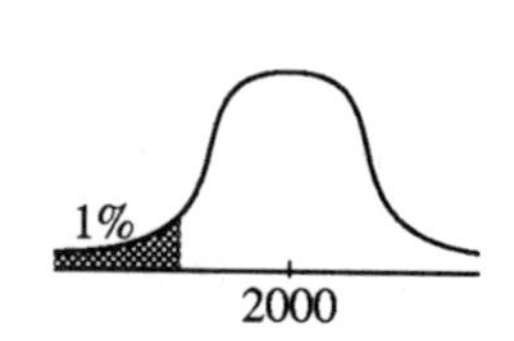

For a one tailed test at 1% significance level, the critical value of z is -2.33 (from normal distribution tables), and here

$$z = \frac{\bar{x}-\mu}{\left(\frac{s}{\sqrt{n}}\right)}$$

$$= \frac{1997-2000}{\left(\frac{12.31}{8}\right)}$$

$$= -1.95.$$

-1.95 is **not** inside the critical region so you conclude that, at 1% level, there is not sufficient evidence to reject H_0.

Exercise 10C

1. Explain, briefly, the roles of a null hypothesis, an alternative hypothesis and a level of significance in a statistical test, referring to your projects where possible.

   A shopkeeper complains that the average weight of chocolate bars of a certain type that he is buying from a wholesaler is less than the stated value of 8.50 g. The shopkeeper weighed 100 bars from a large delivery and found that their weights had a mean of 8.36 g and a standard deviation of 0.72 g. Using a 5% significance level, determine whether or not the shopkeeper is justified in his complaint. State clearly the null and alternative hypotheses that you are using, and express your conclusion in words.

   Obtain, to 2 decimal places, the limits of a 98% confidence interval for the mean weight of the chocolate bars in the shopkeeper's delivery.

2. At an early stage in analysing the marks scored by the large number of candidates in an examination paper, the Examination Board takes a random sample of 250 candidates and finds that the marks, x, of these candidates give $\Sigma x = 11\,872$ and $\Sigma x^2 = 646\,193$. Calculate a 90% confidence interval for the population mean, μ, for this paper.

   Using the figures obtained in this sample, the null hypothesis $\mu = 4.95$ is tested against the alternative hypothesis $\mu < 4.95$ at the α% significance level. Determine the set of values of α for which the null hypothesis is rejected in favour of the alternative hypothesis.

   It is subsequently found that the population mean and standard deviation for the paper are 45.292 and 18.761 respectively. Find the probability of a random sample of size 250 giving a sample mean at least as high as the one found in the sample above.

10.4 Hypothesis testing summary

To summarise, note that:

The null hypothesis H_0 is an assertion that a parameter in a statistical model takes a **particular value.**

The alternative hypothesis H_1 expresses the way in which the value of a parameter may deviate from that specified in the null hypothesis.

Critical region. A value of the test statistic is chosen so that it is very unlikely under H_0 and would be better explained by H_1. If the sample then generates a test statistic in this region defined by the critical value, H_0 will be rejected.

(a) **Two tailed tests**: You do not expect change in any particular direction.

$H_0 : M = k$, a particular value

$H_1 : M \neq k$

Test statistic $z = \dfrac{\bar{x} - \mu}{\left(\dfrac{\sigma}{\sqrt{n}}\right)}$

Testing at the 5% level

The probability of incorrectly rejecting H_0 is 5%.

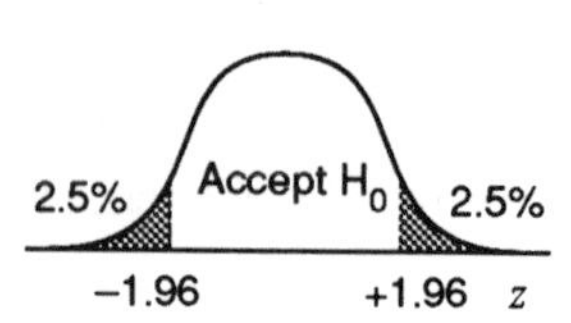

Testing at the 1% level

The probability of incorrectly rejecting H_0 is 1%.

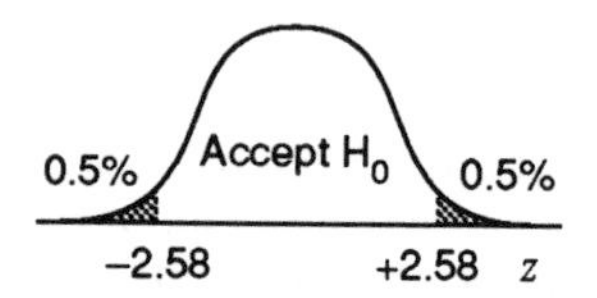

(b) **One tailed tests**: You expect an increase

$H_0 : \mu = K$

$H_1 : \mu > K$

Testing at the 5% level

The probability of incorrectly rejecting H_0 is 5%.

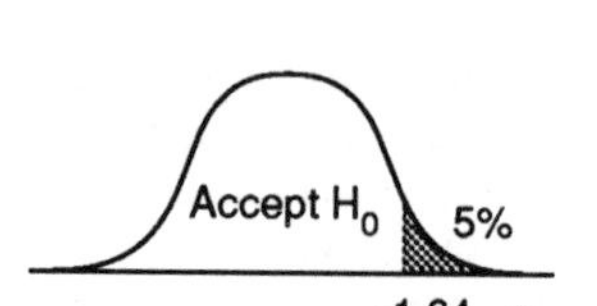

Testing at the 1% level

The probability of incorrectly rejecting H_0 is 1%.

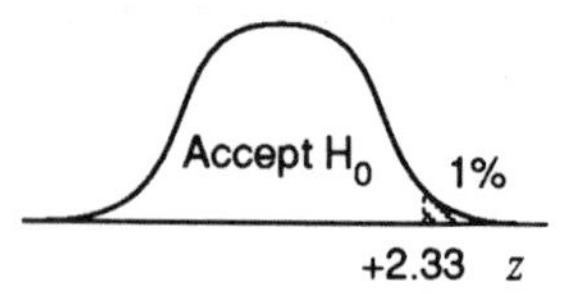

Similarly, if there are grounds for suspecting a decrease,

$H_0 : \mu = K$

$H_1 : \mu < K$

Note

(i) H_0 is the same for every test. It is H_1 which determines the position of the critical region.

(ii) It is always safer to use a two tailed test (unless you have very strong reasons to do otherwise).

Hypothesis testing method

1. Decide which is the variable under investigation.
2. Is it a discrete or a continuous variable?
3. What probability model can you use? (e.g. binomial, normal, uniform)
4. What is the null hypothesis? (H_0)
5. What is the alternative hypothesis? (H_1)
6. Sketch the distribution according to the null hypothesis.
7. Does the alternative hypothesis lead you to look for unusual values of x at one end of the distribution or both? (one or two tailed test?)
8. Is your result significant? (Does it lie in the critical region?)

10.5 Miscellaneous Exercises

1. Nutritionalist working for a babyfood manufacturer wishes to test whether a new variety of orange has a vitamin C content similar to the variety normally used by his company. The original variety of oranges has a mean vitamin C content of 110 milligrams and a standard deviation of 13 mg. His test results are (in mg)

   88, 109, 76, 136, 93, 101, 89, 115, 97, 92,

   106, 114, 109, 91, 94, 85, 87, 117, 105

   What are your conclusions? What assumptions did you need to make?
2. An engineer believes that her newly designed engine will save fuel. A large number of tests on engines of the old variety yielded a mean fuel consumption of 19.5 miles per gallon with standard deviation of 5.2. Fifteen new engines are tested, and give a mean fuel consumption of 21.6 miles per gallon. Is this a significant improvement?
3. A physiotherapist believes that exercise can slow down the ageing process. For the past few years she has been running an exercise class for a group of fourteen individuals whose average age is 50 years. Generally as a person ages, maximum oxygen consumption decreases.

   The national norm for maximum oxygen consumption in 50-year-old individuals is 30 millilitres per kilogram per minute with a standard deviation of 8.6. The mean for the members of the exercise class is 36 millilitres per kilogram per minute. Does the result support the physiotherapist's claim?
4. A coal merchant sells his coal in bags marked 50 kg. He claims that the mean mass is 50 kg with a standard deviation 1 kg. A suspicious weights and measures inspector has 60 of the bags weighed, and finds that their mean mass is 49.6 kg. Are the inspector's suspicions justified?
5. A sample of size 36 is taken from a population having mean μ and standard deviation 9; the sample mean is 47.4.

   Test the hypothesis $H_0 : \mu = 50$ against the alternative $H_1 : \mu < 50$ using the 5% level of significance.
6. A geneticist wishes to test a theory that one ninth of children born to parents both having brown eyes should themselves have blue eyes.

   She examines a random sample of 1200 such children and finds that 113 have blue eyes. What conclusion can she draw about the theory?

11 CHI-SQUARED

Objectives

After studying this chapter you should

- be able to use the χ^2 distribution to test if a set of observations fits an appropriate model;
- know how to calculate the degrees of freedom;
- be able to apply the χ^2 model to contingency tables, including Yates' correction for the 2×2 tables.

11.0 Introduction

The chi-squared test is a particular useful technique for testing whether observed data are representative of a particular distribution. It is widely used in biology, geography and psychology.

Activity 1 How random are your numbers?

Can you make up your own table of random numbers? Write down 100 numbers 'at random' (taking values from 0 to 9). Do this without the use of a calculator, computer or printed random number tables. Draw up a frequency table to see how many times you wrote down each number. (These will be called your **observed** frequencies.)

If your numbers really are random, roughly how many of each do you think there ought to be? (These are referred to these as **expected** frequencies.)

What model are you using for this distribution of expected frequencies?

What assumptions must you make in order to use this model?

Do you think you were able to fulfil those assumptions when you wrote the numbers down?

Can you think of a way to test whether your numbers have a similar frequency distribution to what we would expect for true random numbers?

For analysing data of the sort used in Activity 1 where you are comparing observed with expected values, a chart as shown opposite is a useful way of writing down the data.

	Frequency	
Number	**Observed, O_i**	**Expected, E_i**
1		
2		
3		
4		
.		
.		
.		

11.1 The chi-squared table

For your data in Activity 1, try looking at the differences $O_i - E_i$.

What happens if you total these?

Unfortunately the positive differences and negative differences always cancel each other out and you always have a zero total.

To overcome this problem the differences $O - E$ can be squared. So $\Sigma(O-E)^2$ could form the basis of your 'difference measure'. In this particular example however, each figure has an equal expected frequency, but this will not always be so (when you come to test other models in other situations). The importance assigned to a difference must be related to the size of the expected frequency. A difference of 10 must be more significant if the expected frequency is 20 than if it is 100.

One way of allowing for this is to divide each squared difference by the expected frequency for that category.

Here is an example worked out for you :

	Observed frequency	Expected frequency			
Number	O	E	$O-E$	$(O-E)^2$	$\frac{(O-E)^2}{E}$
0	11	10	1	1	0.1
1	12	10	2	4	0.4
2	8	10	–2	4	0.4
3	14	10	4	16	1.6
4	7	10	–3	9	0.9
5	9	10	–1	1	0.1
6	9	10	–1	1	0.1
7	8	10	–2	4	0.4
8	14	10	4	16	1.6
9	8	10	–2	4	0.4
					6.0

For this set of 100 numbers $\sum \frac{(O-E)^2}{E} = 6$.

But what does this measure tell you?

How can you decide whether the observed frequencies are close to the expected frequencies or really quite different from them?

Firstly, consider what might happen if you tried to test some true random numbers from a random number table.

Would you actually get 10 for each number? The example worked out here did in fact use 100 random numbers from a table and not a fictitious set made up by someone taking part in the experiment.

Each time you take a sample of 100 random numbers you will get a slightly different distribution and it would certainly be surprising to find one with **all** the observed frequencies equal to 10. So, in fact, each different sample of 100 true random numbers will give a different value for $\sum \frac{(O-E)^2}{E}$.

The distribution of $\sum \frac{(O-E)^2}{E}$ is very close to the theoretical distribution known as χ^2 (or chi-squared). In fact, there is a family of χ^2 distributions, each with a different shape depending on the number of **degrees of freedom** denoted by υ (pronounced 'new').

The distribution in this case is denoted by χ^2_υ.

For any χ^2 distribution, the number of degrees of freedom shows the number of independent free choices which can be made in allocating values to the expected frequencies. In these examples, there are ten expected frequencies (one for each of the numbers 0 to 9). However, as the total frequency must equal 100, only nine of the expected frequences can vary independently and the tenth one must take whatever value is required to fulfil that 'constraint'. To calculate the number of degrees of freedom

$$\upsilon = \text{number of classes or groups} - \text{number of constraints.}$$

Here there are ten classes and one constraint so

$$\begin{aligned}\upsilon &= 10-1\\ &= 9.\end{aligned}$$

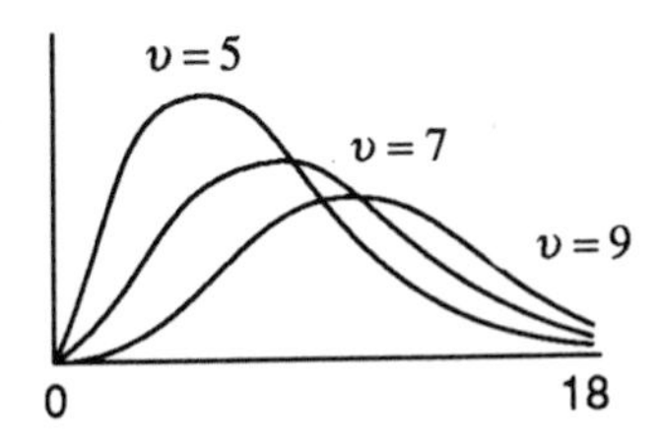

The shape of the χ^2_υ distribution is different for each value of υ and the distribution function is very complicated. The mean of χ^2_υ is υ and the variance is 2υ. The distribution is positively skewed except for large values of υ in which it becomes approximately symmetrical.

Significance testing

The set of random numbers shown in the table on page 204 generated a value of χ^2 equal to 6. You can see where this value comes in the χ^2 distribution with 9 degrees of freedom.

A high value of χ^2 implies a poor fit between the observed and expected frequencies, so the right hand end of the distribution is used for most hypothesis testing.

From χ^2 tables you find that only 5% of all samples of true random numbers will give a value of χ_9^2 greater than 16.92. This is called the **critical value** of χ^2 at 5%. If the **calculated value** of χ^2 from

$$\chi^2 = \sum \frac{(O_i - E_i)^2}{E_i}$$

is less than 16.92, it would support the view that the numbers are random. If not, you would expect that the numbers are not truly random.

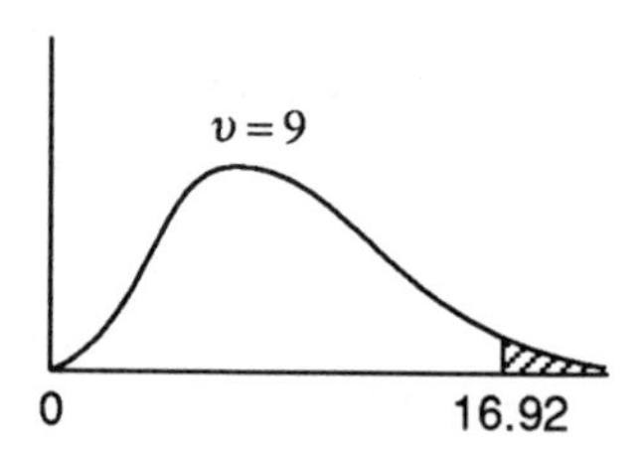

Only 5% of samples of true random numbers give results here

What do you conclude from the example above, where $\chi^2 = 6$?

Activity 2

What happens when you test your made up 'random' numbers? Is their distribution close to what you would expect for true random numbers?

A summary of the critical values for χ^2 at 5% and 1% is given opposite for degrees of freedom $v = 1, 2, \ldots, 10$.

Degree of freedom, v	χ^2 5%	χ^2 1%
1	3.84	6.64
2	5.99	9.21
3	7.82	11.34
4	9.49	13.25
5	11.07	15.09
6	12.59	16.81
7	14.07	18.48
8	15.51	20.09
9	16.92	21.67
10	18.31	23.21

Example

Nadir is testing an octahedral die to see if it is unbiased. The results are given in the table below.

Score	1	2	3	4	5	6	7	8
Frequency	7	10	11	9	12	10	14	7

Test the hypothesis that the die is fair.

Solution

Using χ^2, the number of degrees of freedom is $8-1=7$, so at the 5% significance level the critical value of χ^2 is 14.07. As before, a table of values is drawn up, the expected frequencies being based on a uniform distribution which gives

$$\text{frequency for each result} = \frac{1}{8}(7+10+11+9+12+10+14+7)=10.$$

O	E	$O-E$	$(O-E)^2$	$\frac{(O-E)^2}{E}$
7	10	–3	9	0.9
10	10	0	0	0
11	10	1	1	0.1
9	10	–1	1	0.1
12	10	2	4	0.4
10	10	0	0	0
14	10	4	16	1.6
7	10	–3	9	0.9
				4.0

The calculated value of χ^2 is 4.0. This is well within the critical value, so Nadir could conclude that there is evidence to support the hypothesis that the die is fair.

Exercise 11A

1. Nicki made a tetrahedral die using card and then tested it to see whether it was fair. She got the following scores:

Score	1	2	3	4
Frequency	12	15	19	22

Does the die seem fair?

2. Joe has a die which has faces numbered from 1 to 6. He got the following scores:

Score	1	2	3	4	5	6
Frequency	17	20	29	20	18	16

He thinks that the die may be biased.

What do you think?

3. The table below shows the number of pupils absent on particular days in the week.

Day	M	Tu	W	Th	F
Number	125	88	85	94	108

Find the expected frequencies if it is assumed that the number of absentees is independent of the day of the week.

Test, at 5% level, whether the differences in observed and expected frequencies are significant.

11.2 Contingency tables

In many situations, individuals are classified according to two sets of attributes, and you may wish to investigate the dependency between these attributes. This is dealt with by using a contingency table and the χ^2 distribution.

2×2 contingency tables

The method of approach is illustrated in the example below.

Example

Some years ago a polytechnic decided to require all entrants to a science course to study a non-science subject for one year. In the first year all of the scheme entrants were given the choice of studying French or Russian. The numbers of students of each sex choosing each language are shown in the following table.

	French	Russian
Male	39	16
Female	21	14

Use a χ^2 test (including Yates' correction) at the 5% significance level to test whether choice of language is independent of sex.

Solution

The **observed** frequencies are given in the 2×2 contingency table.

	French	Russian	Total
Male	39	16	55
Female	21	14	35
Total	60	30	90

The null hypothesis is, as usual,

H_0 : there is no relationship between choice of language and sex

and so the alternative hypothesis is

H_1 : the choice of language is dependent on sex.

Assuming the null hypothesis, you need to calculate the expected frequency. For example,

$$P\text{ (student is male)} = \frac{55}{90}$$

$$P\text{ (student is female)} = \frac{60}{90}$$

Since these two events are independent under H_0,

$$P\text{ (student is male and chooses French)} = \frac{55}{90} \times \frac{60}{90},$$

and, since there are 90 students,

$$\text{expected frequency (for male and French)} = \frac{55}{90} \times \frac{60}{90} \times 90$$

$$= \frac{55 \times 60}{90}$$

$$= 36.67.$$

There is no need to go through this procedure each time since it can be calculated directly from

$$\text{Expected frequency} = \frac{\text{(row total) (column total)}}{\text{(grand total)}}$$

In fact, for a 2×2 table only one of these calculations is needed.

The row and column totals can be used to find the other expected values. For example,

$$\text{Expected frequency (for female and French)} = 60 - 36.67$$

$$= 23.33.$$

In this way, the table of expected frequency is as shown below.

	French	Russian	Total
Male	36.67	18.33	55
Female	23.33	11.67	35
Total	60	30	90

Since there is only one expected frequency needed in order to find the rest, the

$$\text{degree of freedom, } \upsilon = 1$$

But, for $\upsilon = 1$, you have to use Yates' continuity correction which evaluates

$$\chi^2_{calc} = \sum_{i-1}^{4} \frac{(|O_i - E_i| - 0.5)^2}{E_i}.$$

From tables, the critical χ^2 at 5% level is given by 3.84. Hence H_0 is rejected if $\chi^2_{calc} > 3.84$. Now

O_i	E_i	$\|O_i - E_i\|$	$\frac{(\|O_i - E_i\| - 0.5)^2}{E_i}$
39	36.67	2.33	0.091
16	18.33	2.33	0.183
21	23.33	2.33	0.144
14	11.67	2.33	0.287

$\chi^2_{calc} = 0.705 < 3.84$,

the critical χ^2 value. Hence you can conclude that there is no evidence to reject H_0; i.e. choice of subject and sex are independent.

Why are all the values in the $|O_i - E_i|$ column the same?

$h \times k$ contingency tables (h rows, k columns)

This is illustrated with an extension to the previous question, which also illustrates the convention that any entry with expected frequency of 5 or less should be eliminated by combining classes together.

Example

Following the example above, the choice of non-science subjects has now been widened and the current figures are as follows

	French	Poetry	Russian	Sculpture
Male	2	8	15	10
Female	10	17	21	37

Use a χ^2 test at the 5% significance level to test whether choice of subject is independent of sex. In applying the test you should combine French with another subject. Explain why this is necessary and the reasons for your choice of subject.

Solution

This is a 2×4 contingency table of **observed** values.

	French	Poetry	Russian	Sculpture	Total
Male	2	8	15	10	35
Female	10	17	21	37	85
Total	12	25	36	47	120

The **expected** frequency for 'male' and 'French' is

$$\frac{12 \times 35}{120} = 3.5.$$

Since this is less than 5, French should be combined with another subject, and the obvious choice is Russian since this is also a language.

Combining the French and Russian together gives

	Fr / Rus	Poetry	Sculpture	Total
Male	17	8	10	35
Female	31	17	37	85
Total	48	25	47	120

As before, H_0 : sex and choice of subject are independent

H_1 : sex and choices of subject are dependent.

The number of degrees of freedom is 2, since determining just 2 expected values will be sufficient to find the rest.

Note that, in general, for an $h \times k$ contingency table

$$\text{No. of degrees of freedom} = (h-1) \times (k-1)$$

(In the example above, $h = 2$, $k = 3$, giving the number of degrees of freedom as $(2-1) \times (3-1) = 1 \times 2 = 2$.) Thus, the critical χ^2 value is 5.99.

The expected frequency for 'male' and 'French and Russian' is is

$$\frac{35 \times 48}{120} = 14.00$$

and for 'male' and 'poetry' is

$$\frac{35 \times 25}{120} = 7.29.$$

The rest of the values can now be calculated from the row and column tables to give the following expected frequences

	Fr / Rus	Poetry	Sculpture	Total
Male	14.00	7.29	13.71	35
Female	34.00	17.71	33.29	85
Total	48	25	47	

and the calculated χ^2 is given by

O_i	E_i	$\lvert O_i - E_i \rvert$	$\frac{(O_i - E_i)^2}{E_i}$
17	14.00	3.00	0.643
8	7.29	0.71	0.069
10	13.71	3.71	1.004
31	34.00	3.00	0.265
17	17.71	0.71	0.028
37	33.29	3.71	0.413

$$\chi^2_{calc} = 2.422 < 5.99$$

the critical value. So you conclude again that there is no dependence between sex and choice of subject.

11.3 Miscellaneous Exercises

1. During an investigation into visits to a Health Centre, interest is focused on the social class of those attending the surgery.

   The table below shows the number of patients attending the surgery together with the population of the whole area covered by the Health Centre, each categorised by social class.

Social Class	I	II	III	IV	V
Patients	28	63	188	173	48
Population	5	500	1600	1200	500

   Use a χ^2 test, at the 5% level of significance, to decide whether or not these results indicate that those attending the surgery are a representative sample of the whole area with respect to social class.

   As part of the same investigation, the following table was constructed showing the reason for the patients' visits to the surgery, again categorised by social class.

	Social Class				
Reason	I	II	III	IV	V
Minor physical	10	21	98	91	27
Major physical	7	17	49	40	15
Mental & other	11	25	41	42	6

   Is there significant evidence to conclude that the reason for the patients' visits to the surgery is independent of their social class?

   Use a 5% level of significance.

   Give an interpretation of your results. (AEB)

2. (a) In a survey on transport, electors from three different areas of a large city were asked whether they would prefer money to be spent on general road improvement or on improving public transport. The replies are shown in the following contingency table.

	Area		
	A	B	C
Road improvement preferred	78	46	24
Public transport preferred	22	34	36

Use a χ^2 test at the 1% significance level to test whether the proportion favouring expenditure on general road improvement is independent of the area.

(b) The same electors were also asked whether they had access to a private car for their personal use. The numbers who did were 70, 40 and 15 respectively in the areas A, B and C respectively and of these 61, 30 and 10 respectively favoured general road improvements.

Construct BUT DO NOT ANALYSE two contingency tables, one for those with access to private cars and one for those without such access.

Given that the value of $\sum \frac{(O-E)^2}{E}$ is 4.21 for the first of these tables and 4.88 for the second of these tables, test in each case, at the 5% significance level whether the proportion favouring general road improvements is independent of area.

(c) Examine your results in (a) and (b) and give an explanation of any apparent inconsistency. (AEB)

3. A hospital employs a number of visiting surgeons to undertake particular operations. If complications occur during or after the operation the patient has to be transferred to a larger hospital nearby where the required back up facilities are available.

A hospital administrator, worried by the effects of this on costs, examines the records of three surgeons. Surgeon A had 6 out of her last 47 patients transferred, surgeon B, 4 out of his last 72 patients and surgeon C, 14 out of his last 41. Form the data into a 2×3 contingency table and test, at the 5% significance level, whether the proportion transferred is independent of the surgeon.

The administrator decides to offer as many operations as possible to surgeon B. Explain why and suggest what further information you would need before deciding whether the administrator's decision was based on valid evidence. (AEB)

4. A group of students studying A-level Statistics was set a paper, to be attempted under examination conditions, containing four questions requiring the use of the χ^2 distribution. The following table shows the type of question and the number of students who obtained good (14 or more out of 20) and bad (fewer than 14 out of 20) marks.

	Type of question			
	Contingency table	Binomial fit	Normal fit	Poisson fit
Good mark	25	12	12	11
Bad mark	4	11	3	12

(a) Test at the 5% significance level whether the mark obtained (by the students who attempted the question) is associated with the type of question.

(b) Under some circumstances it is necessary to combine classes in order to carry out a test. If it had been necessary to combine the binomial fit question with any other question, which question would you have combined it with and why?

(c) Given that a total of 30 students sat the paper, test, at the 5% significance level, whether the number of students attempting a particular question is associated with the type of question.

(d) Compare the difficulty and popularity of the different types of question in the light of your answers to (a) and (b). (AEB)

5. (a) The number of books borrowed from a library during a certain week were 518 on Monday, 431 on Tuesday, 485 on Wednesday, 443 on Thursday and 523 on Friday.

Is there any evidence that the number of books borrowed varies between the five days of the week? Use a 1% level of significance.

Interpret fully your conclusions.

(b) Analysis of the rate of turnover of employees by a personnel manager produced the following table showing the length of stay of 200 people who left the company for other employment.

	Length of employment (years)		
Grade	0-2	2-5	>5
Managerial	4	11	6
Skilled	32	28	21
Unskilled	25	23	50

Using a 1% level of significance, analyse this information and state fully the conclusions from your analysis. (AEB)

12 CORRELATION AND REGRESSION

Objectives

After studying this chapter you should

- be able to investigate the strength and direction of a relationship between two variables by collecting measurements and using suitable statistical analysis;
- be able to evaluate and interpret the product moment correlation coefficient and Spearman's correlation coefficient;
- be able to find the equations of regression lines and use them where appropriate.

12.0 Introduction

Is a child's height at two years old related to her later adult height? Is it true that people aged over twenty have slower reaction times than those under twenty? Does a connection exist between a person's weight and the size of his feet?

In this chapter you will see how to quantify answers to questions of the type above, based on observed data.

12.1 Ideas for data collection

Undertake at least one of the three activities below. You will need your data for further analysis later in this chapter.

Activity 1

Collect a random sample of twenty stones from a beach, gravel driveway or other site. For each stone measure its

(i) maximum dimension
(ii) minimum dimension
(iii) weight.

Does there appear to be a connection between (i) and (ii), (i) and (iii), or (ii) and (iii)?

Activity 2

Measure the heights and weights of a random sample of 15 students of the same sex. Is there any apparent relationship between the two variable?

Would you expect the same relationship (if any) to exist between the heights and weights of the opposite sex?

Activity 3

Collect a dozen volunteers and time them running a forty metre straight sprint. Ask them to do two long jumps each and record the better one. (Measure the jump from the point of take-off rather than any board.)

Is there a connection between the times and distances recorded?

12.2 Studying results

The data below gives the marks obtained by 10 pupils taking Maths and Physics tests.

Pupil	A	B	C	D	E	F	G	H	I	J
Maths mark (out of 30) x	20	23	8	29	14	11	11	20	17	17
Physics mark (out of 40) y	30	35	21	33	33	26	22	31	33	36

Is there a connection between the marks gained by ten pupils, A, B, C ..., J in Maths and Physics tests?

A starting point would be to plot the marks as a scatter diagram.

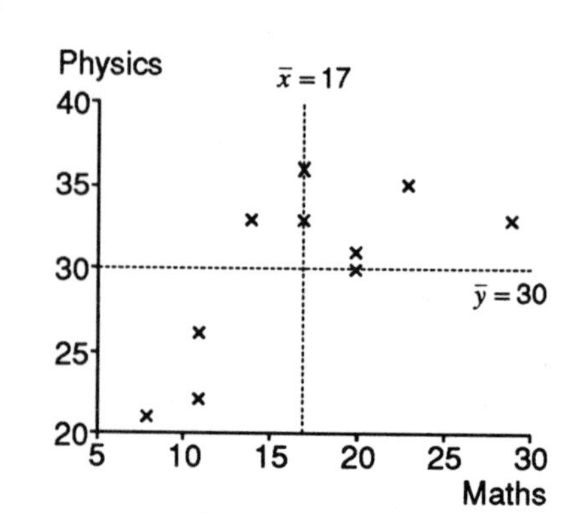

The areas in the bottom right and top left of the graph are largely vacant so there is a tendency for the points to run from bottom left to top right.

Calculating the means,

$$\bar{x}=\frac{170}{10}=17$$

and $$\bar{y}=\frac{300}{10}=30$$

and using them to divide the graph into four shows this clearly.

The problem is to find a way to measure how strong this tendency is.

Covariance

An attempt to quantify the tendency to go from bottom left to top right is to evaluate the expression

$$s_{xy} = \frac{1}{n}\sum_{i=1}^{n}(x_i - \bar{x})(y_i - \bar{y})$$

which is known as the **covariance** and denoted by $\text{cov}(X,Y)$ or s_{xy}. For shorthand it is normally written as

$$\frac{1}{n}\Sigma\,(x - \bar{x})(y - \bar{y})$$

where the summation over i is assumed.

The points in the top right have x and y values greater than $\bar{x}$ and $\bar{y}$ respectively, so $x - \bar{x}$ and $y - \bar{y}$ are both positive and so is the product $(x - \bar{x})(y - \bar{y})$.

Those in the bottom left have values less than $\bar{x}$ and $\bar{y}$, so $x - \bar{x}$ and $y - \bar{y}$ are both negative and again the product $(x - \bar{x})(y - \bar{y})$ is positive.

Points in the other two areas have one of $x - \bar{x}$ and $y - \bar{y}$ positive and the other negative, so $(x - \bar{x})(y - \bar{y})$ is negative.

The $\frac{1}{n}$ factor accounts for the fact that the number of points will affect the value of the covariance.

In the example above, most of the points give positive values of $(x - \bar{x})(y - \bar{y})$.

There is another form of the expression for covariance which is easier to use in calculations.

$$\begin{aligned}\frac{1}{n}\Sigma(x - \bar{x})(y - \bar{y}) &= \frac{1}{n}\Sigma(xy - \bar{x}y - x\bar{y} + \bar{x}\bar{y}) \\ &= \frac{1}{n}(\Sigma xy - \Sigma\bar{x}y - \Sigma\, x\bar{y} + \Sigma\bar{x}\bar{y})\end{aligned}$$

$$= \frac{1}{n}\left(\Sigma xy - \bar{x}\Sigma y - \bar{y}\Sigma x + n\overline{xy}\right)$$

$$= \frac{1}{n}\left(\Sigma xy - \bar{x}ny - \bar{y}nx + n\overline{xy}\right) \qquad \text{since} \quad \bar{y} = \frac{\Sigma y}{n}, \quad \bar{x} = \frac{\Sigma x}{n}$$

$$= \frac{1}{n}\left(\Sigma xy - n\overline{xy}\right).$$

Thus

$$\boxed{\frac{1}{n}\Sigma\,(x-\bar{x})(y-\bar{y}) = \frac{1}{n}\Sigma xy - \overline{xy}}$$

The right hand side is quicker to evaluate. For the example on page 216, this form of the expression is usually used when calculating covariance.

$$s_{xy} = \frac{1}{10}\Sigma xy - 17\times 30$$

$$= \frac{1}{10}\times 5313 - 510$$

$$= 21.3$$

(Σxy is a function available on calculators with LR mode.)

The fact that $s_{xy} > 0$ indicates that the points follow a trend with a positive slope. The size of the number, however, conveys little as it can easily be altered by a change of scale.

The following examples show this.

Example

Find the covariance for the following data.

(a)

Height (m) x	1.60	1.64	1.71
Weight (kg) y	53	57	60

(b)

Height (cm) x	160	164	171
Weight (kg) y	53	57	60

Solution

(a) $$s_{xy} = \frac{1}{3}\times 280.88 - \frac{170}{3}\times\frac{4.95}{3}$$

$$= 0.12\dot{6}$$

(b) $s_{xy} = \frac{1}{3} \times 28088 - \frac{170}{3} \times \frac{495}{3}$

$= 12.\dot{6}$

You can, of course, get quite different values by measuring in pounds and inches or kg and feet, etc. They will all be positive but their sizes will not convey useful information.

Activity 4

Find the covariance for the data you collected in any of the first three activities.

12.3 Pearson's product moment correlation coefficient

Dividing $(x - \bar{x})$ by the standard deviation s_x gives the distance of each x value above or below the mean as so many standard deviations. For the example on height and weight above, the standard deviations in m and cm are related, with the second being one hundred times the first, so

$$\frac{x - \bar{x}}{s_x}$$

will give the same answer regardless of the units or scale involved. The quantity

$$\frac{1}{n} \sum \left(\frac{x - \bar{x}}{s_x} \right) \left(\frac{y - \bar{y}}{s_y} \right)$$

can therefore be relied on to produce a value with more meaning than the covariance.

Since

$$\frac{1}{n} \sum \left(\frac{x - \bar{x}}{s_x} \right) \left(\frac{y - \bar{y}}{s_y} \right) = \frac{\frac{1}{n} \Sigma xy - \overline{xy}}{s_x s_y}$$

and the latter is easier to evaluate, **Pearson's product moment correlation coefficient** is often given as

$$r = \frac{\frac{1}{n}\Sigma xy - \overline{x}\overline{y}}{s_x s_y}$$

where $s_x = \sqrt{\frac{1}{n}\Sigma x^2 - \overline{x}^2}$ and $s_y = \sqrt{\frac{1}{n}\Sigma y^2 - \overline{y}^2}$.

(Note that r is a function given on calculators with LR mode.)

Returning to the example in Section 12.2:

Pupil	A	B	C	D	E	F	G	H	I	J
Maths mark (out of 30) x	20	23	8	29	14	11	11	20	17	17
Physics mark (out of 40) y	30	35	21	33	33	26	22	31	33	36

$$r = \frac{\frac{1}{10} \times 5313 - 17 \times 30}{s_x \times s_y}$$

$$s_x = \sqrt{\frac{1}{10} \times 3250 - 17^2} = \sqrt{36} = 6$$

$$s_y = \sqrt{\frac{1}{10} \times 9250 - 30^2} = \sqrt{25} = 5$$

$$\Rightarrow \quad r = \frac{531.3 - 510}{6 \times 5}$$

$$= 0.71$$

The value of r gives a measure of how close the points are to lying on a straight line. It is always true that

$$\boxed{-1 \le r \le 1}$$

and $r = 1$ indicates that all the points lie exactly on a straight line with positive gradient, while $r = -1$ gives the same information with line having a negative gradient, and $r = 0$ tells us that there is no connection at all between the two sets of data.

The sketches opposite indicate these and in between cases.

(Note that s_{xy} is not a calculator key, but its value may be checked by $r \times s_x \times s_y$ which are all available.)

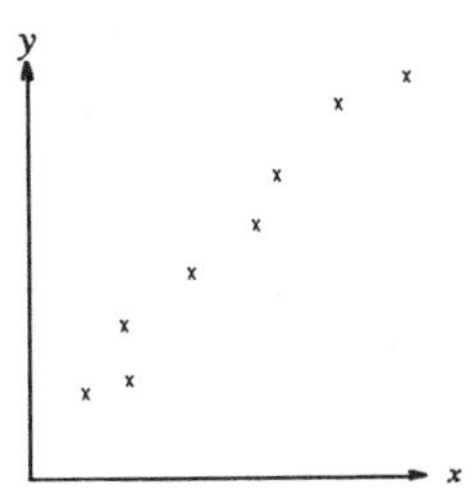
Strong positive correlation

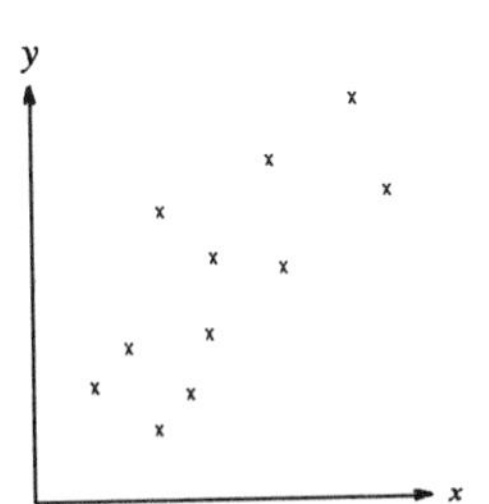
Weak positive correlation

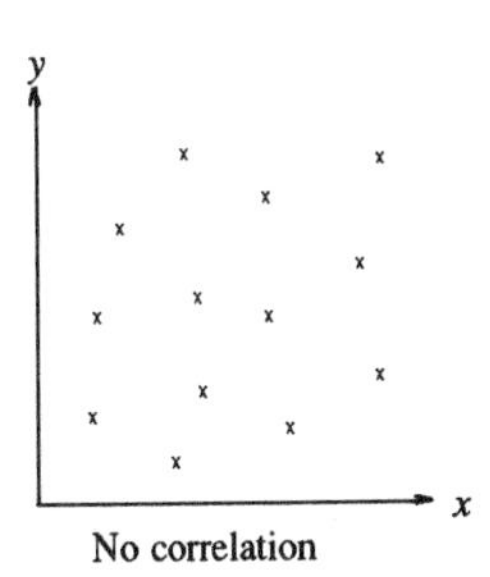
No correlation

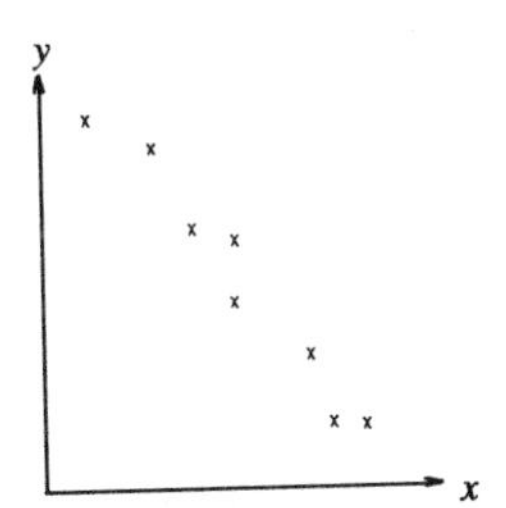
Strong negative correlation

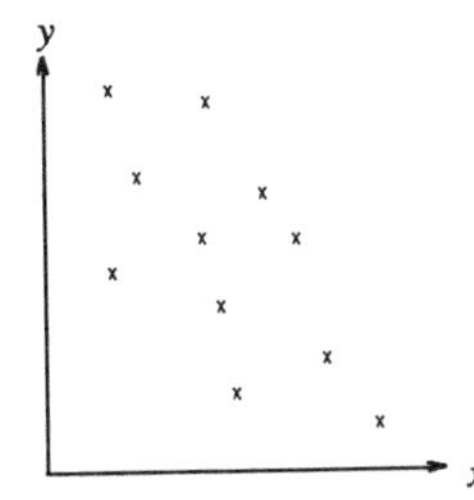
Weak negative correlation

The significance of r

With only **two** pairs of values it is unlikely that they will lie on the same horizontal or vertical line, giving a correlation coefficient of zero but any other arrangement will produce a value of r equal to plus or minus one, depending on whether the line through them has a positive or negative gradient. With **six** points, however, the fact that they lie on, or close to, a straight line becomes much more significant.

The following table, showing critical values at 5% significance level, gives some indication of how likely some values of the correlation coefficient are. For example, for $n = 5$, $r = 0.878$ means that there is only a 5% chance of getting a result of 0.878 or greater if there is **no** correlation betwen the variables. Such a value, therefore, indicates the likely existence of a relationship between the variables.

(no.of pairs) n	r
3	0.997
4	0.950
5	0.878
6	0.811
7	0.754
8	0.707
9	0.666
10	0.632

More detailed tables of critical values are given in the Appendix for a range of significant limits and values of n. Their calculation relies on the data being drawn from joint normal distributions, so using them in other circumstances cannot provide an accurate assessment of significance.

Example

A group of twelve children participated in a psychological study designed to assess the relationship, if any, between age, x years, and average total sleep time (ATST), y minutes. To obtain a measure for ATST, recordings were taken on each child on five consecutive nights and then averaged. The results obtained are shown in the table.

Child	Age (x years)	ATST (y minutes)
A	4.4	586
B	6.7	565
C	10.5	515
D	9.6	532
E	12.4	478
F	5.5	560
G	11.1	493
H	8.6	533
I	14.0	575
J	10.1	490
K	7.2	530
L	7.9	515

$\Sigma x = 108$ $\Sigma y = 6372$ $\Sigma x^2 = 1060.1$ $\Sigma y^2 = 3396942$ $\Sigma xy = 56825.4$

Calculate the value of the product moment correlation coefficient between x and y. Assess the statistical significance of your value and interpret your results.

Solution

(a) Use the formula

$$s_{xy} = \frac{1}{n}\Sigma xy - \overline{xy}$$

when $\bar{x} = \frac{108}{12} = 9$ and $\bar{y} = \frac{6372}{12} = 531$.

Thus $s_{xy} = \frac{1}{12}(56825.4) - 9 \times 531 = -43.55$

Also $s_x = \sqrt{\frac{1}{12} \times 1060.1 - 9^2} \approx 2.7096$

$$s_y = \sqrt{\frac{1}{12} \times 3396942 - 531^2} \approx 33.4290$$

Hence $r = \dfrac{-43.55}{2.7096 \times 33.4290} \approx -0.481$

This indicates weak negative correlation. But to apply a significance test, the null and alternative hypotheses need to be defined:

$$H_0: r = 0$$
$$H_1: r \neq 0$$

significance level : 5% (two tailed).

Using the earlier table of critical values, for $n = 12$,

$$r_{crit} = \pm 0.576$$

Since $r = -0.481$, there is insufficient evidence to reject the null hypothesis.

Limitations of correlation

You should note that

(1) r is a measure of **linear** relationship only. There may be an exact connection between the two variables but if it is not a straight line r is no help. It is well worth studying the scatter diagram carefully to see if a non-linear relationship may exist. Perhaps studying x and $\ln y$ may provide an answer but this is only one possibility.

(2) Correlation does not imply **causality**. A survey of pupils in a primary school may well show that there is a strong correlation between those with the biggest left feet and those who are best at mental arithmetic. However it is unlikely that a policy of 'left foot stretching' will lead to improved scores. It is possible that the oldest children have the biggest left feet and are also best at mental arithmetic.

(3) An unusual or freak result may have a strong effect on the value of r. What value of r would you expect if point P were omitted in the scatter diagram opposite?

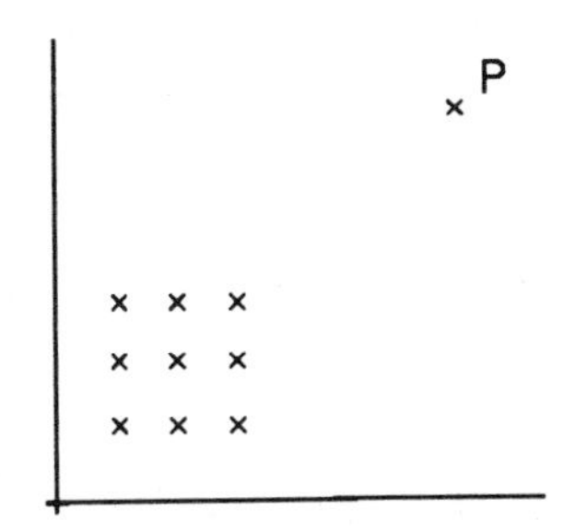

Exercise 12A

1. For each of the following sets of data

   (a) draw a scatter diagram

   (b) calculate the product moment correlation coefficient.

(i)

x	1	3	6	10	12
y	5	13	25	41	49

(ii)

x	1	3	5	7	9
y	44	34	24	14	4

(iii)

x	1	1	3	5	5
y	5	1	3	1	5

(iv)

x	1	3	6	9	11
y	12	28	37	28	12

2. (a) Calculate the value of r for the random variables X and Y using the following values

x	11	17	26
y	23	18	19

(b) The random variable Z is converted to Y by the equation $Z = \frac{Y}{10} + 3$.

x	11	17	26
Z			

Complete the table above and evaluate r for X and Z.

(c) State the value of r for Y and Z.

3. The diameter of the longest lichens growing on gravestones were measured.

Age of gravestone x (years)	Diameter of lichen y (mm)
9	2
18	3
20	4
31	20
44	22
52	41
53	35
61	22
63	28
63	32
64	35
64	41
114	51
141	52

Draw a scatter diagram to show the data.

Calculate the values of $\bar{x}$ and $\bar{y}$ and show these as vertical and horizontal lines. Which three points are the odd ones out?

Find the values of s_x, s_y and r.

4. In a biology experiment a number of cultures were grown in the laboratory. The numbers of bacteria, in millions, and their ages, in days, are given below.

Age (x)	1	2	3	4	5	6	7	8
No.of bacteria (y)	34	106	135	181	192	231	268	300

(i) Plot these on a scatter diagram with the x-axis having a scale up to 15 days and the y-axis up to 410 millions. Calculate the value of r and comment on your results.

(ii) Some late readings were taken and are given below.

x	13	14	15
y	400	403	405

Add these points to your graph and describe what they show.

5. A metal rod was gradually heated and its length, L, was measured at various temperatures, T.

Temperature (°C)	15	20	25	30	35	40
Length (cm)	100	103.8	106.1	112	116.1	119.9

Draw a scatter diagram to show the data and evaluate r. (Plot L against T.)

Do you suspect a major inaccuracy in any of the recorded values? If so, discard any you consider untrustworthy and find the new value of r.

12.4 Spearman's rank correlation coefficient

Two judges at a fete placed the ten entries for the 'best fruit cakes' competition in order as follows (1 denotes first, etc.)

Entry	A	B	C	D	E	F	G	H	I	J
Judge 1 (x)	2	9	1	3	10	4	6	8	5	7
Judge 2 (y)	6	9	2	1	8	4	3	10	7	5

No actual marks like 73/100 have been awarded in this case where only ranks exist.

Is there a linear relationship between the rankings produced by the two judges?

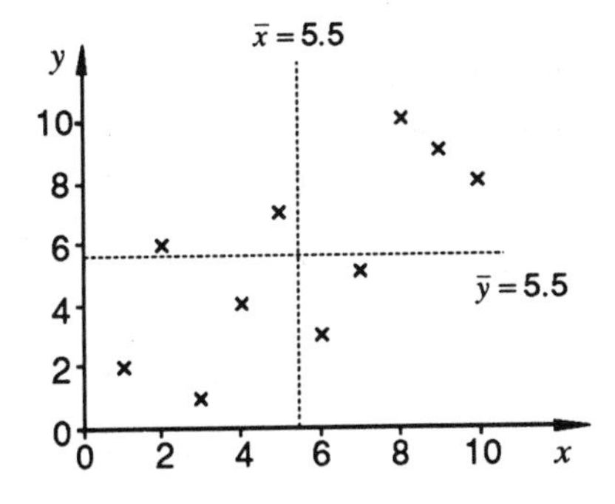

Spearman's rank correlation coefficient answers this question by simply using the ranks as data and in the product moment coerrelation coefficient, r, and denoting it r_s. Again a scatter diagram may be drawn and the presence of the points plotted in, or very near, the top right and bottom left areas indicates a positive correlation.

Spearman's rank correlation coefficient,

$$r_s = \frac{\frac{1}{10}\Sigma xy - \bar{x}\bar{y}}{s_x s_y}$$

where $\bar{x} = \bar{y} = \dfrac{55}{10} = 5.5$

$$s_x = s_y = \sqrt{\frac{385}{10} - 5.5^2} = \sqrt{8.25}$$

and $\Sigma xy = 2\times 6 + 9\times 9 + \ \dots\ + 7\times 5\ = 362$

$$\Rightarrow \quad r_s = \frac{\frac{1}{10}\times 362 - 5.5^2}{\sqrt{8.25}\sqrt{8.25}}$$

$$= \frac{36.2 - 30.25}{8.25}$$

$$= 0.721$$

(The significance tables for r should certainly not be used here as the ranks definitely do not come from normal distributions.)

It can be shown that, when there are no tied ranks,

$$\frac{\frac{1}{n}\Sigma xy - \bar{x}\bar{y}}{s_x s_y} = 1 - \frac{6\Sigma d^2}{n(n^2-1)}$$

and so

$$\boxed{r_s = 1 - \frac{6\Sigma d^2}{n(n^2-1)}}$$

where $d = x - y$, is the difference in ranking.

For the example just considered

Entry	A	B	C	D	E	F	G	H	I	J
Judge 1	2	9	1	3	10	4	6	8	5	7
Judge 2	6	9	2	1	8	4	3	10	7	5
$\lvert d\rvert$	4	0	1	2	2	0	3	2	2	2
d^2	16	0	1	4	4	0	9	4	4	4

So $\qquad \Sigma d^2 = 16+0+1+ \ \dots \ +4 = 46$

$$\Rightarrow \quad r_s = 1 - \frac{6\times 46}{10(100-1)}$$

$$= 1 - \frac{6\times 46}{10\times 99}$$

$$= \frac{119}{165}$$

$$\approx 0.721 \text{ to 3 decimal places.}$$

As with the product moment correlation coefficient, Spearman's correlation coefficient also obeys

$$-1 \le r_s \le 1$$

where $r = 1$ corresponds to perfect positive correlation and $r = -1$ to perfect negative correlation.

The definition of the formula from the product moment correlation coefficient will not be given here but you will see in the following Activity how it can be deduced.

Activity 5

You can verify Spearman's formula by first assuming that

$$r_s = 1 - K\Sigma d^2$$

where K is a constant for each value of n.

(a) Show that $r_s = 1$ for perfect positive correlation.

(b) Use the fact that $r_s = -1$ for perfect negative correlation to complete the table below.

n	2	3	4	5	6	7	8
K							

(For example, for $n = 4$, perfect negative correlation corresponds to

1	2	3	4
4	3	2	1)

Check that these values agree with the Spearman's formula, that is

$$K = \frac{6}{n(n^2-1)}.$$

Significance of r_s

If the tables of significance for r cannot be used here, you can still assess the importance of the value by noting that the formula

$$r_s = 1 - \frac{6\Sigma d^2}{n(n^2-1)}$$

contains the term Σd^2. Tables giving the significance of its possible value for various values of n are given in the Appendix.

So at 5% significant level, the hypotheses are defined by

$$H_0: r_s = 0$$

$$H_1: r_s \neq 0 \quad \text{(two tailed)}$$

and, with $n = 10$, the tables show that

$$p(|r_s| > 0.6485) = 0.05$$

So for a two tailed test, you should reject H_0 since in the example on page 226, $r_s = 0.721$, and accept H_1, the alternative hypothesis, which says that there is significant correlation.

You can test for positive correlation, by using the hypothesis

$$H_0: r_s = 0$$

$$H_1: r_s > 0 \quad \text{(one tailed)}$$

At 5% level, and with $n = 10$ as before,

$$p(r_s > 0.5636) = 0.05$$

and since $0.721 > 0.5636$, again H_0 is rejected. You accept the alternative hypothesis that there is significant positive correlation.

Tied ranks

The form $r_s = 1 - \frac{6\Sigma d^2}{n(n^2-1)}$ does not give the correct value for r_s when there are **tied** ranks, but as long as you do not have too many ties, the inaccuracies are negligible and the use of this equation allows the table of significance for Σd^2 to be employed.

Example

Find the value of n for the following data

Ranks x	1	2 =	2 =	5	4	6	7	8
Ranks y	1	3	4	2	5	6 =	6 =	6 =

Solution

Those tied in the x rankings are given a value of $\frac{2+3}{2} = 2\frac{1}{2}$ and those tied in y are allocated $\frac{6+7+8}{3} = 7$. (In general, each tie is given the mean of the places that would have been occupied if a strict order had been produced.) The table, therefore, becomes

Ranks x	1	$2\frac{1}{2}$	$2\frac{1}{2}$	5	4	6	7	8
Ranks y	1	3	4	2	5	7	7	7
$\lvert d \rvert$	0	$\frac{1}{2}$	$1\frac{1}{2}$	3	1	1	0	1
d^2	0	$\frac{1}{4}$	$2\frac{1}{4}$	9	1	1	0	1

Hence $\quad \Sigma d^2 = 14.5$

and
$$r_s = 1 - \frac{6 \times 14.5}{8(64-1)}$$
$$= 0.827$$

and again this is a significant result. That is, you would conclude that there is positive correlation.

Exercise 12B

1.

Item	A	B	C	D	E	F
Ranks (x)	1 =	1 =	1 =	4 =	4 =	4 =
Ranks (y)	1	2	3 =	3 =	5	6

Use both

(i) $r_s = 1 - \dfrac{6\Sigma d^2}{n(n^2-1)}$ and (ii) $r_s = \dfrac{\frac{1}{n}\Sigma xy - \overline{x}\overline{y}}{s_x s_y}$

to evaluate Spearman's rank correlation coefficient for the two sets of rankings given.

Comment on your results.

2. The performances of the six fastest male sprinters in a school were noted in their winter cross-country race. The details are shown in the table.

	A	B	C	D	E	F
Sprint ranking	1	2	3	4	5	6
Position in cross-country	70	31	4	32	12	17

Give each athlete a rank for cross-country and evaluate r_s. Comment on the significance of your result.

3. At an agricultural show 10 Shetland sheep were ranked by a qualified judge and by a trainee judge. Their rankings are shown in the table.

Qualified Judge 1	2	3	4	5	6	7	8	9	10
Trainee Judge 1	2	5	6	7	8	10	4	3	9

Calculate a rank correlation coefficient for these data. Is this result significant at the 5% level?

4. Five sacks of coal, A, B, C, D and E have different weights, with A being heavier than B, B being heavier than C, and so on. A weight lifter ranks the sacks (heaviest first) in the order A, D, B, E, C. Calculate a coefficient of rank correlation between the weight lifter's ranking and the true ranking of the weights of the sacks.

5. A company is to replace its fleet of cars. Eight possible models are considered and the transport manager is asked to rank them, from 1 to 8, in order of preference. A saleswoman is asked to use each type of car for a week and grade them according to their suitability for the job (A - very suitable to E - unsuitable). The price is also recorded.

Model	Transport manager's ranking	Saleswoman's grade	Price (£10's)
S	5	B	611
T	1	B+	811
U	7	D-	591
V	2	C	792
W	8	B+	520
X	6	D	573
Y	4	C+	683
Z	3	A-	716

(a) Calculate Spearman's rank correlation coefficient between

(i) price and transport manager's rankings,

(ii) price and saleswoman's grades.

(b) Based on the results in (a) state, giving a reason, whether it would be necessary to use all three different methods of assessing the cars.

Ideas for data collection

Activity 6

Use a strong elastic band or spring as a simple weighing machine. Carefully hang weights from it and record its length for each one.

Activity 7

Mount two metres of toy railway track on flexible board. Raise one end and record the distance travelled by a railway truck for each different height.

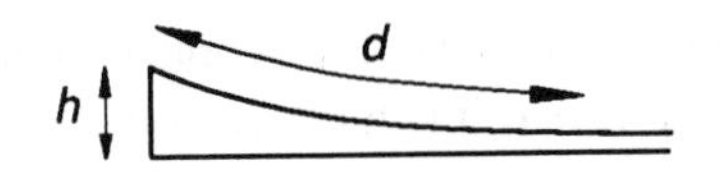

Activity 8

Run water into a container on a set of scales. The water should flow in at as steady a rate as possible from just above the level of the container. Record the time taken for the scales to show different masses. e.g.

Mass (g) x	200	250	300	350	400
Time (secs) y					

12.5 Linear regression

In linear regression you start by looking at a set of points to see if there is a relationship between them and if there is you proceed to establish it in such a way that further points may be deduced from it with the minimum possible error. That is, start with points, proceed to a line and regress to points again.

Here are some results for the elastic band experiment suggested in Activity 5.

Mass g (x)	50	100	150	200	250	300	350	400
Length mm (y)	37	48	60	71	80	90	102	109

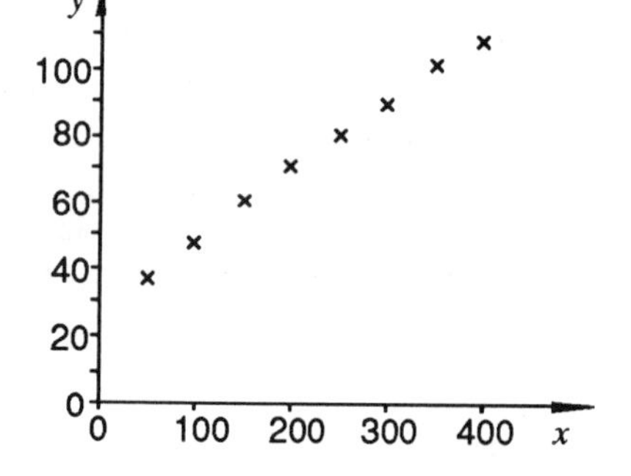

In the diagram opposite, the points lie very close to a straight line and the value of r is 0.999.

Activity 9

Find the value of

(a) r, the product moment correlation coefficient.

(b) r_s, Spearman's rank correlation coefficient.

Comment on their values.

Having decided that the points follow a straight line, with some small variations due to errors in measurement, changes in the environment etc, the problem is to find the line which best fits the data.

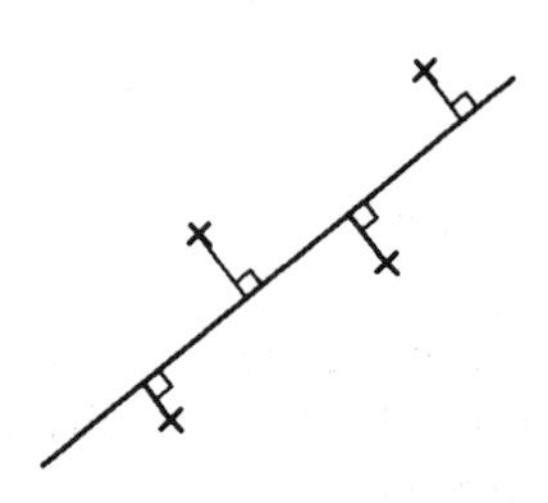

It may seem natural to try to find the line so that the points' distances from it have as small a total as possible. However, since the line will need to produce values of y for given values of x (or vice versa) it is more sensible to seek to produce a line so that any distances in the y direction, and therefore any errors in predicting y given x, should be a minimum.

If the line is to be used to predict values of y based on known values of x it is called the 'y on x' line and its equation is determined by making $d_1^2 + d_2^2 + \ldots = \Sigma d^2$ as small as possible. The equation of this line can be shown to be

$$\boxed{y - \bar{y} = \frac{s_{xy}}{s_x^2}(x - \bar{x})}$$

and for this line $\Sigma d^2 = n s y^2 (1 - r^2)$. You will notice that when $r = \pm 1$ (i.e. the points lie exactly on a straight line) then $\Sigma d^2 = 0$ as would be expected. The procedure used to obtain the equation is called the **method of least squares** and the 'd's are often referred to as the **residuals**. The gradient is called the **regression coefficient.**

For the elastic band example,

$$\bar{x} = \frac{1800}{8} = 225, \quad \bar{y} = \frac{597}{8} = 74.625$$

$$s_{xy} = \frac{156150}{8} - 225 \times 74.625 = 2728.125$$

$$s_x^2 = \frac{510000}{8} - 225^2 = 13125$$

$$\Rightarrow \quad y - 74.625 = \frac{2728.125}{13125}(x - 225)$$

$$\Rightarrow \quad \boxed{y = 0.208x + 27.857}$$

The values of 0.208 and 27.857 represent the gradient of the line and its intercept on the y-axis and are available directly from a calculator with LR mode. The gradient has units mm/g and tells us how much extension would be caused by the addition of 1 extra gram to the suspended mass. This line can now be used to find values of y given values of x.

Example

What length would you expect the elastic band to be if a weight of

(a) 375 g (b) 1 kg

was suspended by it?

Solution

(a) $\hat{y} = 0.208 \times 375 + 27.857$

$= 105.9$ mm

(The ^ above the y indicates that this is an estimate, however accurate. Calculators with LR mode usually have a $\hat{y}$ function giving the answer directly.)

(b) $\hat{y} = 0.208 \times 1000 + 27.857$

$= 235.9$

The first of these answers is an example of **interpolating,** (that is 'putting between' known values) and is quite trustworthy. The latter, though, is a case of **extrapolating** (that is 'putting beyond' known values) and may be wildly inaccurate. The elastic may well break under the action of the 1 kg mass!

The mass x is known as the **independent** or **exploratory variable** and is controlled by the experimenter. The length y is called the **dependent** or **response variable** and is less accurate. For any fixed value of x used repeatedly the resulting readings for y will form a normal distribution.

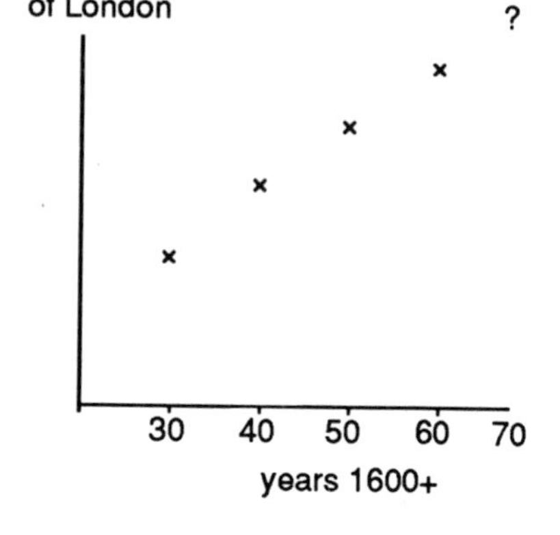

It may be tempting to extrapolate in the example illustrated opposite, and modern day planners have to do just that, but the plague and the great fire of 1666 would be guaranteed to sabotage any attempt in this case.

Any estimates outside the range of the data are dangerous and the further away they are the less trust can be placed in them.

Estimates of n based on given values of y may be obtained from the line but since it was constructed to minimise errors in the y direction it was not designed for this use, so answers are bound to be unreliable.

Drawing the line

Looking at the equation

$$y - \bar{y} = \frac{s_{xy}}{s_x^{\,2}} \ (x - \bar{x})$$

we can see that $x = \bar{x}$, $y = \bar{y}$ satisfies it so $(\bar{x}, \bar{y})$ will always be a point on the line. To find a couple more points to enable you to draw the line use the $\hat{y}$ values with the two x values at the ends of the given set of values.

So, for the elastic band example,

$$x = 50 \Rightarrow \hat{y} = 38.3$$

$$x = 400 \Rightarrow \hat{y} = 111.$$

Other forms of the equation

Since $$y - \bar{y} = \frac{s_{xy}}{s_x^{\,2}}(x - \bar{x})$$

$$\Rightarrow \quad \frac{y - \bar{y}}{s_y} = \frac{s_{xy}}{s_x s_y}\left(\frac{x - \bar{x}}{s_x}\right)$$

$$\Rightarrow \quad \boxed{\frac{y - \bar{y}}{s_y} = r\left(\frac{x - \bar{x}}{s_x}\right)}$$

Also $$\frac{s_{xy}}{s_x^{\,2}} = \frac{\frac{1}{n}\Sigma(x - \bar{x})(y - \bar{y})}{\frac{1}{n}\Sigma(x - \bar{x})^2}$$

$$= \frac{\Sigma(x - \bar{x})(y - \bar{y})}{\Sigma(x - \bar{x})^2}$$

so $$y - \bar{y} = \hat{\beta}(x - \bar{x})$$

where $$\hat{\beta} = \frac{\Sigma(x - \bar{x})(y - \bar{y})}{\Sigma(x - \bar{x})^2} = \frac{n\sum xy - \sum x \sum y}{n\sum x^2 - (\sum x)^2}$$

Exercise 12C

1. A student counted the number of words in an essay she had written, recording the total every 10 lines.

No. of lines (x)	10	20	30	40	50	60	70	80
No. of words (y)	75	136	210	291	368	441	519	588

Find the formula to convert lines to words. How many words (approximately) has she written if she writes

(a) 65 lines (b) 100 lines (c) 1000 lines?

Are you happy with all these estimates?

2. Eight test areas were given different concentrations of a new fertiliser and the resulting crop was weighed.

Concentration g/L (x)	1	2	3	4	5	6	7	8
Weight of crop kg(y)	7	11.1	14	16.2	20	23.9	27	29

Draw a scatter diagram to show the data. Calculate the equation of the regression line y on x and show it on your diagram.

What increase in weight of crop might be expected from raising the concentration of fertiliser by 1 g/L?

3. An experiment was carried out to investigate variation of solubility of chemical X in water. The quantities in kg that dissolved in 1 litre at various temperatures are show in the table.

Temp. °C (y)	15	20	25	30	35	50	70
Mass of X (x)	2.1	2.6	2.9	3.3	4.0	5.1	7.0

Draw a scatter diagram to show the data. Calculate the equation of the regression line of y on x. Draw the line and plot the point $(\bar{x}, \bar{y})$ on your diagram. What quantity might be expected to dissolve at 42° C? Find the quantity that your equation indicates would dissolve at −10° C and comment on your answer.

Calculate the sum of the squares of the residuals and comment on your result.

12.6 Bivariate distributions

In many situations it may not be possible to control either variable.

Example

In a decathlon held over two days the following performances were recorded in the high jump and long jump. All distances are in metres.

Competitor	A	B	C	D	E	F	G
High jump x	1.90	1.85	1.96	1.88	1.88	Abs	1.92
Long jump y	6.22	6.24	6.50	6.36	6.32	6.44	Abs

What performances might have been expected from F in the high jump and G in the long jump if they had competed?

Solution

To estimate G's performance in the long jump we use the y on x line.

Now $$y - \bar{y} = \frac{s_{xy}}{s_x^{\,2}}(x - \bar{x})$$

and using competitors A to E,

$$\bar{y} = \frac{31.64}{5} = 6.328, \qquad \bar{x} = \frac{9.47}{5} = 1.894$$

Also $$s_{xy} = \frac{1}{5} \times 59.9404 - 6.328 \times 1.894 = 0.002848$$

$$s_x^{\,2} = \frac{1}{5} \times 17.9429 - 1.894^2 = 0.001344$$

$$\Rightarrow \quad y - 6.328 = \frac{0.002848}{0.001344}(x - 1.894)$$

$$\Rightarrow \quad y = 2.119x + 2.315$$

Thus $x = 1.92$ gives $\hat{y} = 2.119 \times 1.92 + 2.315 = 6.38\text{m}$

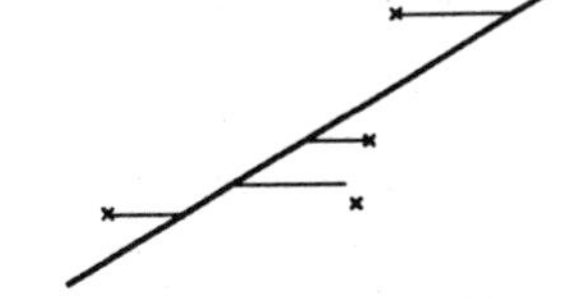

Now to estimate F's high jump accurately we need a line for which the sum of the horizontal distances from it is a minimum.

This is the x on y line and its equation is

$$x - \bar{x} = \frac{s_{xy}}{s_y^{\,2}}(y - \bar{y})$$

$$s_y^{\,2} = \frac{1}{5} \times 200.268 - 6.328^2 = 0.010016$$

$$\Rightarrow \quad x - 1.894 = \frac{0.002848}{0.010016}\;(y - 6.328)$$

$$\Rightarrow \quad x = 0.284y + 0.095.$$

Now $\quad y = 6.44 \quad \Rightarrow \quad \hat{x} = 0.284 \times 6.44 + 0.095 = 1.92$ m

(To use all the functions available in LR mode the coordinates can be typed in with the pairs reversed)

Notice that

$$x = 1.92 \quad \Rightarrow \quad \hat{y} = 6.38$$
$$y = 6.44 \quad \Rightarrow \quad \hat{x} = 1.92$$

Might we have expected $x = 1.92 \Rightarrow \hat{y} = 6.44$**?**

Not really as the two predictions are made from different lines.

y on x and x on y lines

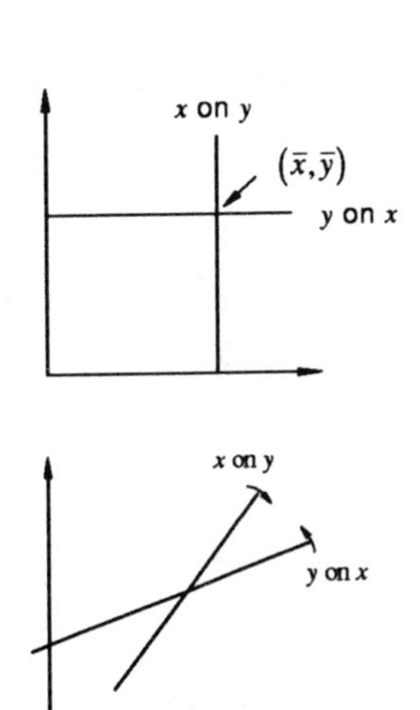

When $r = 0$ the y on x line is horizontal as can be seen from the form

$$\frac{y - \bar{y}}{s_y} = r\left(\frac{x - \bar{x}}{s_x}\right).$$

Similarly the x on y line is vertical as it has the form

$$\frac{x - \bar{x}}{s_x} = r\left(\frac{y - \bar{y}}{s_y}\right).$$

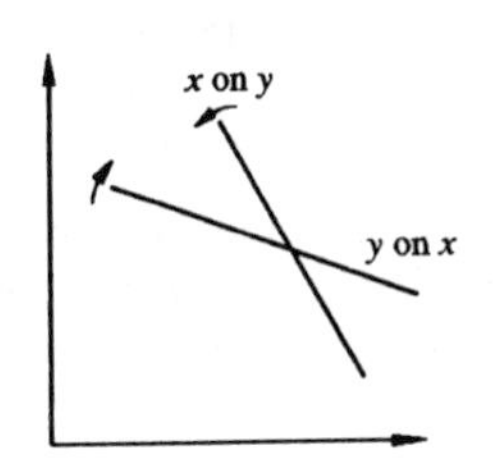

As r increases from zero the lines rotate about their point of intersection until they coincide when $r = 1$ as a line with positive gradient.

As r decreases from zero they turn about $(\bar{x}, \bar{y})$ until they meet as a single line with negative gradient when $r = -1$.

Exercise 12D

1. In an investigation into prediction using the stars and planets a celebrated astrologist Horace Cope predicted the ages at which thirteen young people would first marry. The complete data, of predicted and actual ages at first marriage, are now available and are summarised in the table.

Person	A	B	C	D	E	F	G	H	I	J	K	L	M
Predicted age x (years)	24	30	28	36	20	22	31	28	21	29	40	25	27
Actual age y (years)	23	31	28	35	20	25	45	30	22	27	40	27	26

(a) Draw a scatter diagram of these data.

(b) Calculate the equation of the regression line of y on x and draw this line on the scatter diagram.

(c) Comment upon the results obtained, particularly in view of the data for person G. What further action would you suggest?

(AEB)

2. The experimental data below were obtained by measuring the horizontal distance y cm, rolled by an object released from the point P on a plane inclined at $\theta°$ to the horizontal, as shown in the diagram.

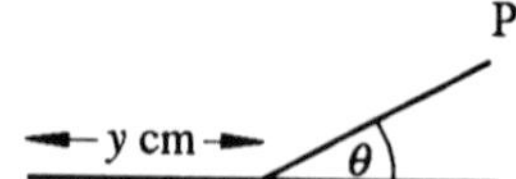

Distance y	Angle $\theta°$
44	8.0
132	25.0
152	31.5
87	17.5
104	20.0
91	10.5
142	28.5
76	14.5

$\Sigma y = 828$, $\Sigma y\theta = 18147$

$\Sigma \theta = 155.5$, $\Sigma \theta^2 = 3520.25$

(a) Illustrate the data by a scatter diagram.

(b) Calculate the equation of the regression line of distance on angle and draw this line on the scatter diagram.

(c) It later emerged that one of the points was obtained using a different object.

Suggest which point this was.

(d) Estimate the distance the original object would roll if released at an angle of (i) 12°, (ii) 40°. Discuss the uncertainty of each of these estimates.

3. The variables H and T are known to be linearly related. Fifty pairs of experimental observations of the two variables gave the following results:

$\Sigma H = 83.4$, $\Sigma T = 402.0$,

$\Sigma HT = 680.2$, $\Sigma H^2 = 384.6$

$\Sigma T^2 = 3238.2$.

Obtain the regression equation from which one can estimate H when T has the value 7.8 and give, to 1 decimal place, the value of this estimate.

4. Students were asked to estimate the centres of the two 10 cm lines shown below.

(i)

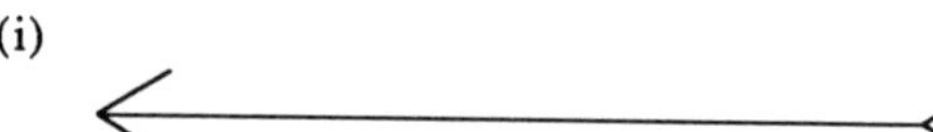

(ii)

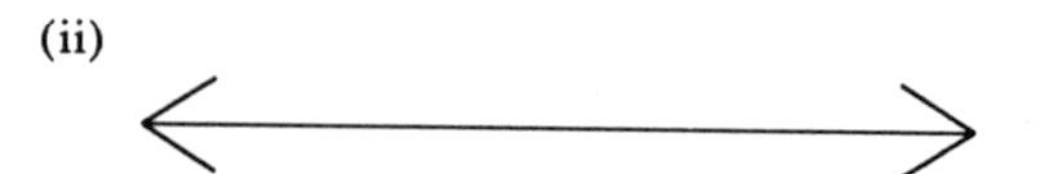

Their errors are shown in the following table with '–' indicating an error to the left of the centre (all in mm).

Error on (i) x	1	4	7	6	2	0	1	4
Error on (ii) y	0	1	2	2	−1	0	−1	3

Draw a scatter diagram to show the data. Calculate the equations of the regression lines y on x and x on y.

Draw both lines and plot $(\bar{x}, \bar{y})$ on your diagram.

Estimate

(a) y when $x = 5$

(b) x when $y = 1$

12.7 Miscellaneous Exercises

1. The yield of a batch process in the chemical industry is known to be approximately linearly related to the temperature, at least over a limited range of temeratures. Two measurements of the yield are made at each of eight temperatures, within this range, with the following results:

Temperature (°C) x	180	190	200	210	220	230	240	250
Yield (tonnes) y	136.2	147.5	153.0	161.7	176.6	194.2	194.3	196.5
	136.9	145.1	155.9	167.8	164.4	183.0	175.5	219.3

$\Sigma x = 1720 \qquad \Sigma x^2 = 374000$

(a) Plot the data on scatter diagram.

(b) For each temerature, calculate the mean of the two yields. Calculate the equation of the regression line of this mean yield on temperature. Draw the regression line on your scatter diagram.

(c) Predict, from the regression line, the yield of a batch at each of the following temperatures:

(i) 175 (ii) 185 (iii) 300

Discuss the amount of uncertainty in each of your three predictions.

(d) In order to improve predictions of the mean yield at various temperatures in the range 180 to 250 it is decided to take a further eight measurements of yield. Recommend, giving a reason, the temperatures at which these measurements could be carried out.
(AEB)

2. Some children were asked to eat a variety of sweets and classify each one on the following scale:

strongly dislike/dislike/neutral/like/like very much.

This was then converted to a numerical scale 0, 1, 2, 3, 4 with 0 representing 'strongly dislike'. A similar method produced a score on the scale 0, 1, 2, 3 for the sweetness of each sweet assessed by each child (the sweeter the sweet the higher the score). The following frequency distribution resulted

		Liking				
		0	1	2	3	4
Sweetness	0	5	2	0	0	0
	1	3	14	16	9	0
	2	8	22	42	29	37
	3	3	4	36	58	64

(a) Calculate the product moment correlation coefficient for these data. Comment briefly on the data and on the correlation coefficient.

(b) A child was asked to rank 7 sweets according to preference and sweetness with the following results:

	Ranks						
Sweet	A	B	C	D	E	F	G
Preference	3	4	1	2	6	5	7
Sweetness	2	3	4	1	5	6	7

Calculate Spearman's rank correlation coefficient for these data.

(c) It is suggested that the product moment correlation coefficient should be calculated for (b) and Spearman's rank correlation coefficient for (a). Comment on this suggestion.
(AEB)

3. A lecturer gave a group of students an assignment consisting of two questions. The following table summarises the number of numerical errors made on each question by the group of students.

	Errors on Question 1 (x)				
Errors on Question 2 (y)	0	1	2	3	4
0				4	3
1				4	5
2		5	7	5	2
3	1	4	3	4	

(a) Find the product moment correlation coefficient between x and y.

(b) Give a written interpretation of your answer.

The scores on each question for a random sample of 8 of the group are as shown below.

Student	1	2	3	4	5	6	7	8
Question 1	42	68	32	84	71	55	55	70
Question 2	39	75	43	79	83	65	62	68

(c) Calculate the Spearman rank correlation coefficient between the scores on the two questions.

(d) Give an interpretation of your result. (AEB)

4. Sets of china are individually packed to customers' requirements. The packaging manager introduces a new procedure in which each packer is responsible for all stages of an order from its initial receipt to final despatch. In order to be able to estimate the time to pack particular orders, he recorded the time taken by a particular packer to complete his first 11 sets packed by the new system. The data are in order of packing.

No. of items in set x	40	21	62	49	21	30
Time in min. to complete packaging, y	545	370	525	440	315	285

No. of items in set x	10	57	48	20	38
Time in min. to complete packaging, y	220	410	360	285	320

(a) Draw a scatter diagram of the data. Label the points from 1 to 11 according to the order of packing.

(b) Calculate the regression line of 'time' on 'number of items' and draw it on your scatter diagram. Comment on the pattern revealed and suggest why it has occurred.

(c) The regression line for the last 6 points only is $y = 188 + 3.70x$. Draw this line on your scatter diagram.

(d) The packaging manager estimated that the next order, which consisted of 44 items, would take 406.31 minutes. Comment on this estimate and the method by which you think it was made.

Make your own estimate of the packaging time for this order and explain why you think it is better than the packaging manager's.

(AEB)

5. A headteacher wished to investigate the relationship between coursework marks for GCSE and marks for internal school examinations. She asked the Head of English and the Head of Science to provide some data. The Head of English reported that the marks for his four best students were as follows:

Exam mark, x	84	79	89	92
Coursework mark, y	86	85	81	91

(a) Calculate the product moment correlation coefficient for these marks.

(b) The Head of Science reported that he had asked every teacher in the school to supply him with full details of all marks. Not everyone had cooperated and some subjects used letter grades instead of marks. However, he had converted all information received into a score from 1 to 3 (the better the grade the higher the score). He produced the following frequency distribution:

		Examination score x		
		1	2	3
Course work score y	1	940	570	310
	2	630	1030	720
	3	290	480	1910

Calculate the product moment correlation coefficient for these marks.

(c) Comment on the two sets of data provided and their appropriateness to the investigation. What advice would you give the headteacher if she were to carry out a similar exercise next year? (AEB)

6. In an attempt to increase the yield (kg/h) of an industrial process a technician varies the percentage of a certain additive used, while keeping all other conditions as constant as posible. The results are shown below.

Yield y	127.6	130.2	132.7	133.6	133.9	133.8	133.3	131.9
% additive x	2.5	3.0	3.5	4.0	4.5	5.0	5.5	6.0

You may assume that $\sum x = 34$ $\sum y = 1057$ $\sum xy = 4504.55$ $\sum x^2 = 155$.

(a) Draw a scatter diagram of the data.

(b) Calculate the equation of the regression line of yield on percentage additive and draw it on the scatter diagram.

The technician now varies the temperature (°C) while keeping other conditions as constant as possible and obtains the following results:

Yield y	127.6	128.7	130.4	131.2	133.6
Temperature t	70	75	80	85	90

He calculates (correctly) that the regression line is $y = 107.1 + 0.29t$.

(c) Draw a scatter diagram of these data together with the regression line.

(d) The technician reports as follows, 'The regression coefficient of yield on percentage additive is larger than that of yield on temperature, hence the most effective way of increasing the yield is to make the percentage additive as large as possible, within reason'.

Criticise the report and make your own recommendations on how to achieve the maximum yield. (AEB)

7. An instrument panel is being designed to control a complex industrial process. It will be necessary to use both hands independently to operate the panel. To help with the design it was decided to time a number of operators, each carrying out the same task once with the left hand and once with the right hand.

The times, in seconds, were as follows:

Operator	A	B	C	D	E	F	G	H	I	J	K
l.h., x	49	58	63	42	27	55	39	33	72	66	50
r.h., y	34	37	49	27	49	40	66	21	64	42	37

You may assume that

$\sum x = 554$ $\sum x^2 = 29902$ $\sum y = 466$

$\sum y^2 = 21682$ $\sum xy = 24053$

(a) Plot a scatter diagram of the data.

(b) Calculate the product moment correlation coefficient between the two variables and comment on this value.

(c) Further investigation revealed that two of the operators were left handed. State, giving a reason, which you think these were. Omitting their two results, calculate Spearman's rank correlation coefficient and comment on this value.

(d) What can you say about the relationship between the times to carry out the task with left and right hands? (AEB)

8. An electric fire was switched on in a cold room and the temperature of the room was noted at intervals.

Time in minutes, from switching on the fire, x								
0	5	10	15	20	25	30	35	40
Temperature, °C, y								
0.4	1.5	3.4	5.5	7.7	9.7	11.7	13.5	15.4

You may assume that

$\sum x = 180$ $\sum y = 68.8$ $\sum xy = 1960$

$\sum x^2 = 5100$

(a) Plot the data on a scatter diagram.

(b) Calculate the regression line $y = a + bx$ and draw it on your scatter diagram.

(c) Predict the temperature 60 minutes from switching on the fire. Why should this prediction be treated with caution?

(d) Starting from the equation of the regression line $y = a + bx$, derive the equation of the regression line of

(i) y on t where y is temperature in °C (as above) and t is time in hours.

(ii) z on x where z is temperature in K and x is time in minutes (as above).

(A temperature in °C is converted to K by adding 273, e.g. 10°C → 283 K)

(e) Explain why, in (b), the line $y = a + bx$ was calculated rather than $x = a' + b'x$. If, instead of the temperature being measured at 5 minute intervals, the time for the room to reach predetermined temperatures (e.g. 1, 4, 7, 10, 13°C) had been observed what would the appropriate calculation have been? Explain your answer. (AEB)

9. The data in the following table show the length and breadth (in mm) of a group of skulls discovered during an excavation.

Length (x)	165	170	172	176	178	179	182	184	186	190
Breadth (y)	139	141	147	147	149	149	159	145	155	152

(You may assume that $\sum x^2 = 318086$ and $\sum xy = 264582$.)

(a) Calculate the regression lines of length on breadth and breadth on length.

(b) Plot these data on a scatter diagram and draw both your regression lines on your diagram.

(c) State, in symbolic form, the point of intersection of your two lines.

(d) Using in each case the appropriate regression line predict the breadth of a skull of length 185 mm and the length of a skull of breadth 155 mm.

(e) Under what circumstances would your two lines be coincident? (AEB)

10. A small firm negotiates an annual pay rise with each of its twelve employees. In an attempt to simplify the process it is proposed that each employee should be given a score, x, based on his/her level of responsibility. The annual salary will be £$(a+bx)$ and the annual negotiations will only involve the values of a and b. The following table gives last year's salaries (which were generally accepted as fair) and the proposed scores.

Employee	x	Annual salary (£), y
A	10	5750
B	55	17300
C	46	14750
D	27	8200
E	17	6350
F	12	6150
G	85	18800
H	64	14850
I	36	9900
J	40	11000
K	30	9150
L	37	10400

(You may assume that $\Sigma x=459$, $\Sigma x^2=22889$, $\Sigma y=132600$ and $\Sigma xy=6094750$)

(a) Plot the data on a scatter diagram.

(b) Estimate values that could have been used for a and b last year by fitting the regression line $y=a+bx$ to the data. Draw the line on the scatter diagram.

(c) Comment on whether the suggested method is likely to prove reasonably satisfactory in practice.

(d) Without recalculating the regression line find the appropriate values of a and b if every employee were to receive a rise of

(i) £500 per year

(ii) 8%

(iii) 4% plus £300 per year.

(e) Two employees, B and C, had to work away from home for a large part of the year. In the light of this additional information, suggest an improvement to the model. (AEB)

11. The following data show the IQ and the score in an English test of a sample of 10 pupils taken from a mixed ability class.

The English test was marked out of 50 and the range of IQ values for the class was 80 to 140.

Pupil	A	B	C	D	E	F	G	H	I	J
IQ (x)	110	107	127	100	132	130	98	109	114	124
English Score (y)	26	31	37	20	35	34	23	38	31	36

(a) Estimate the product moment correlation coefficient for the class.

(b) What does this coefficient measure?

(c) Outline briefly how other information given in the data of the question might have affected your coefficient.

For two other groups within the class, the teacher assessed each individual in terms of scholastic aptitude and perseverance. A rating scale of 0–100 was used for each assessment and the following table summarises the ratings for one of the groups.

Scholastic aptitude	42	68	32	84	71	55	58	70
Perseverance	39	75	43	79	83	65	62	68

(d) Show that the Spearman rank correlation coefficient between the two sets of ratings for the group is 0.905.

(e) The value of the Spearman rank correlation coefficient between the sets of ratings for the other group is -0.886. Interpret briefly the sign of each of these coefficients.

(f) When these two groups are combined, the value of the Spearman rank correlation coefficient is 0.66. Interpret and explain the effect of this combining on the correlation between scholastic aptitude and perseverance. (AEB)

12. (a) The product moment correlation coefficient between the random variables W and X is 0.71 and between the random variables Y and Z is -0.05.

For each of these pairs of variables, sketch a scatter diagram which might represent the results which gave the correlation coefficients.

(b) The scatter diagram on the next page shows the amounts of the pollutants, nitrogen oxides and carbon monoxide, emitted by the exhausts of 46 vehicles. Both variables are measured in grams of the pollutant per mile driven.

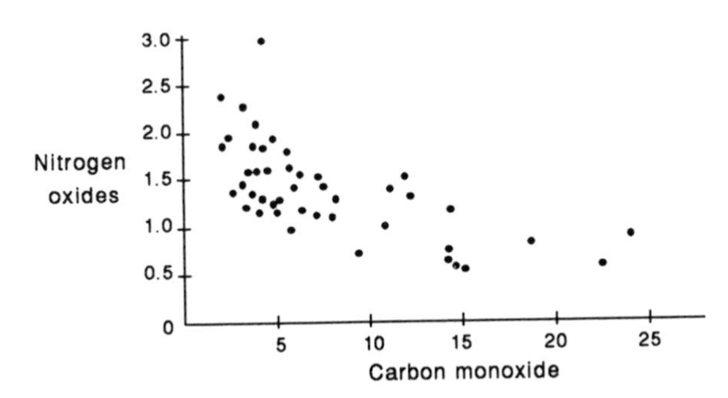

Write down three noticeable features of this scatter diagram.

It has been suggested that,

'If an engine is out of tune, it emits more of all the important pollutants. You can find out how badly a vehicle is polluting the air by measuring any one pollutant. If that value is acceptable, the other emissions will also be acceptable.'

State, giving your reason, whether or not this scatter diagram supports the above suggestion.

(c) When investigating the amount of heat evolved during the hardening of cement, a scientist monitored the amount of heat evolved, Y, in calories/g of cement, and four explanatory variables, X_1, X_2, X_3 and X_4. Based on thirteen observations, the scientist produced the following correlation matrix.

	Y	X_1	X_2	X_3	X_4
Y	1	0.731	0.816	−0.535	−0.821
X_1		1	0.229	r	−0.245
X_2			1	−0.139	−0.973
X_3				1	0.030
X_4					1

The values of X_1 and X_3 are as follows.

x_1	7	1	11	11	7	11	3	1	2	21	1	11	10
x_3	6	15	8	8	6	9	17	22	18	4	23	9	8

Assuming $\Sigma x_1^2 = 1139$ and $\Sigma x_3^2 = 2293$, find r, the product moment correlation coefficient between X_1 and X_3.

Write down two noticeable features of the correlation matrix. (AEB)

ANSWERS

The answers to the questions set in the Exercises are given below. Answers to questions set in some of the Activities are also given where appropriate.

1 PROBABILITY

Exercise 1A

1. $\frac{4}{8}=\frac{1}{2}$

2. $\frac{22}{36}=\frac{11}{18}$

3. $\frac{3}{5}$

4. $\frac{1}{7}$

5. Six ways

6. 16

7. $P(\text{Pierre wins})=\frac{15}{36}=\frac{5}{12}$;

   $P(\text{Julian wins})=\frac{7}{12}$;

   $P(\text{Julian wins})=\frac{1}{2}$

8. $\frac{69}{169}$

9. $\frac{2}{7}$

Exercise 1B

1. (a) $\frac{9}{20}$ (b) 0 (c) 1

2. (a) $\frac{605+17}{\text{total}}=\frac{622}{3272}=0.19$

   (b) 0.14 (c) 0.19 (d) $\frac{17}{447}=0.04$

Exercise 1C

1. $\frac{1}{2}+\frac{1}{3}-\frac{1}{10}=\frac{11}{15}$

2. $\frac{1}{3}+\frac{1}{4}=\frac{7}{12}$ (mutually exclusive)

3. $\frac{11}{30}$

4. (a) neither (b) exclusive (c) exclusive
   (d) exhaustive (e) neither (f) exclusive

Exercise 1D

1. (b) $\frac{3}{8}$

2. $\frac{2}{7}$

3. (a) 0.34 (b) 0.063 (c) 0.19 (d) 0.97
   Choosing all 3 white marbles

Exercise 1E

1. (a) $\frac{1}{17}$ (b) $\frac{13}{204}$ (c) $\frac{13}{51}$

2. (a) $\frac{1}{3}$ (b) $\frac{2}{15}$ (c) $\frac{8}{15}$

3. (a) 0.02 (b) 0.78 (c) 0.76 (d) $\frac{1}{30}$

4. (a) 0.5 (b) 0.35 (c) 0.375 (d) 0.4

Exercise 1F

1. (a) $\frac{4}{52}=\frac{1}{13}$ (b) $\frac{1}{13}$ (c) $\frac{13}{52}=\frac{1}{4}$

   (d) $\frac{3}{12}=\frac{1}{4}$ (e) $\frac{13}{26}=\frac{1}{2}$ (f) $\frac{12}{52}=\frac{3}{13}$

   (g) 0 (h) $\frac{12}{48}=\frac{1}{4}$ (i) $\frac{4}{50}=\frac{2}{25}$

(j) $\frac{1}{2}$ (k) $\frac{12}{13}$ (l) $\frac{25}{50}=\frac{1}{2}$

2. a, b, d

3. $a=\frac{1}{3}$ $b=\frac{2}{3}$ $c=\frac{1}{10}$ $d=\frac{2}{5}$ $e=\frac{1}{4}$ $f=\frac{3}{4}$

$g=\frac{1}{2}$

4. (a) $\frac{1}{2}\times\frac{7}{15}+\frac{1}{2}\times\frac{7}{12}=\frac{21}{40}$ (b) $\frac{\frac{1}{2}\times\frac{7}{12}}{\frac{21}{40}}=\frac{5}{9}$

5. 36, 18, 6;

(a) $\frac{133}{295}$ (b) $\frac{46}{295}$ (c) $\frac{\frac{1}{118}}{\frac{133}{295}}=\frac{5}{266}$

6. (a) $\frac{19}{216}$ (b) $\frac{7}{12}$

7. (a) $\frac{39}{50}$ (b) $\frac{7}{100}$, $\frac{10}{11}$

Miscellaneous Exercises

1. $\frac{13}{18}$

2. (a) $\frac{1}{15}$ (b) $\frac{2}{5}$

3. (a) $\frac{1}{5}$ (b) $\frac{1}{3}$ (c) $\frac{49}{90}$ (d) $\frac{1}{15}$

4. $\frac{5}{16}$

5. $\frac{1}{8}$

8. (a) true $P(A\cap B)=0$ or $P(A|B)=0$
   (b) true $P(A\cup B)=1$
   (c) true $P(B\cup D)=1$
   (d) false $P(A'\cap C')=\frac{14}{16}=\frac{7}{8}$

9. $\frac{11}{30}$

10. (a) $\frac{1}{4}+\frac{2}{5}=\frac{13}{20}$ (b) $1-\frac{13}{20}=\frac{7}{20}$

11. (a) $\frac{1}{2}$ (b) $\frac{3}{4}$ (c) $\frac{5}{6}$ (d) $\frac{5}{12}$ (e) $\frac{4}{9}$

12. (a) 0.43 (b) $\frac{6}{19}$

13. (a) 0.575 (b) 0.1 (c) $\frac{3}{23}$

14. (a) $\frac{416}{720}=\frac{26}{45}$ (b) $\frac{239}{360}$ (c) $\frac{63}{242}$

15. $\frac{1}{4}$, £1, 66 mm (old size), 57 mm (new size)

16. 90°, 108°, 162°

17. $\frac{4}{19}$, yes $\frac{3}{19}$

18. (a) $\frac{19}{27}$ (b) $\frac{6}{19}$

19. (a) $\frac{2}{5}$ (b) $\frac{7}{12}$ $\frac{x}{x+y}$

20. (a) $\frac{1}{81}$ (b) $\frac{80}{81}$

21. (a) $\frac{6}{25}$ (b) $\frac{1}{2}$ (c) $\frac{1}{5}$

22. $\frac{5}{16}$

23. (a) $\frac{98}{125}$ (b) $\frac{4}{49}$

24. (a) $\frac{1}{7}$ (b) $\frac{1}{4}$ (c) $\frac{15}{28}$

25. Yes

26. (a) (i) 0.0429 (ii) 0.142 (iii) 0.1215
    (iv) 0.189 (v) 0.334
    (b) 0.642

27. (a) $\frac{11}{18}$ (b) (i) 0.16 (ii) 0.5 (iii) 0.62

28. (a) (i) 0.4 (ii) 0.224 (iii) 0.944 (iv) 0.35
    (b) (i) C (ii) B (iii) $\overline{A}$
    (c) (i) 0.716 (ii) 0.117

29. (a) 0.012 (b) 0.03 (c) 0.4

30. (a) 0.55 (b) 0.04 (c) 0.006 (d) $\frac{1}{3}$
    (e) 0.024 (f) 0.12 (g) 0.01 (h) 0.55

2 DATA COLLECTION

Miscellaneous Exercises

1. (b) 24, 24, 26, 14, 8, 4
3. Method 1: random but sensible
   Method 2: random
   Method 3: not random
   Method 4: random
   Method 5: systematic but not random
4. 3, 9, 10, 15, 3

3 DESCRIPTIVE STATISTICS

Exercise 3C

2. (b) Cons: 283; Lib/Dem: 169; Other: 14

Exercise 3E

2. (i) 258 (ii) 172; £86

Activity 5

(a) (i) 62 (ii) 82 (b) 37 kg

Exercise 3E

2. (i) 258 (ii) 172
   £73.36

Exercise 3F

2. About 76
3. About 395

Exercise 3G

2. (a) 1.42

Exercise 3I

1. 5.5 min
3. 6100 ; 9400

Exercise 3J

1. Cycling : 3.48 min
   Running: 3.79 min
2. 1941 : 28.37, 6.14
   1961 : 27.09, 6.01
   1989 : 26.81, 5.52
3. Males - employees : 39.7, 8.2
   Males - self-employed : 47.6, 12.6
   Females - employees : 29.2, 12.0
   Females - self-employed : 32.8, 18.5

Miscellaneous Exercises

2. (a) £83.4, £108.1
3. £8235.6 ; £8857.7
4. (a) 736.1, 744.1, 752.3
   (b) mean = 744.2, SD = 14.9
6.

		Mean	SD
1st Year	B	2.73	1.74
	G	2.41	1.67
3rd Year	B	2.73	1.67
	G	2.49	1.67
5th Year	B	2.51	1.65
	G	2.21	1.60

Combined

1st Year	2.57	1.71
3rd Year	2.61	1.67
5th Year	2.36	1.63

7. (b) 19.72, 73.5 (c) 4.40, 7.79, 12.14
10. (i) 52 (ii) 41 (iii) 62
11. (c) (i) 8 (ii) 5-15
12. (a) 14 (b) 4.5 (c) 5.6%
13. (a) 2100 (b) 900 (c) 35%
14. (b) 20.3, 5
15. 27.5, 8.5
16. 2, 3.2, 3.53, 1.985; 0.48, 0.77
17. (b) 128, 40 (d) 137

4 DISCRETE PROBABILITY DISTRIBUTIONS

Exercise 4A

1. (a) discrete (b) continuous (c) discrete
   (d) discrete
2. (a)

Score	2	3	4	5	6	7	8	9	10	11	12
P	$\frac{1}{36}$	$\frac{2}{36}$	$\frac{3}{36}$	$\frac{4}{36}$	$\frac{5}{36}$	$\frac{6}{36}$	$\frac{5}{36}$	$\frac{4}{36}$	$\frac{3}{36}$	$\frac{2}{36}$	$\frac{1}{36}$

   (b) 7

3.

X	0	1	2	3
P	$\frac{1}{8}$	$\frac{3}{8}$	$\frac{3}{8}$	$\frac{1}{8}$

   (c) 1.5

4. (a) $c=\dfrac{1}{20}$ (c) 2.3
5. $x=\dfrac{1}{6}$, $y=\dfrac{1}{12}$

Exercise 4B

1. $V(X)=\frac{35}{12}$, $s\approx 1.71$

2. (a) $V(X)=\frac{35}{6}$, $s\approx 2.42$

   (b) $V(X)=\frac{3}{4}$, $s\approx 0.866$

3.

X	0	1	2	3
P	$\frac{1}{35}$	$\frac{12}{35}$	$\frac{18}{35}$	$\frac{4}{35}$

   (a) $E(X)=\frac{60}{35}$ (b) $V(X)=\frac{24}{49}$, $s\approx 0.700$

Exercise 4C

1. (a) 2 (b) 1

2. (a) $P(S=s)=\frac{1}{10}$ for $s=1,2,\ldots,10$

   (b) $\frac{11}{2}$ (c) $\sqrt{\frac{33}{4}}\approx 2.87$

3. (a) $E(X)=1$; $V(X)=1$

   (b) $E(Y)=0$; $V(Y)=1.2$

4. (a)

X	2	3	4	5	6
P	$\frac{1}{36}$	$\frac{4}{36}$	$\frac{10}{36}$	$\frac{12}{36}$	$\frac{9}{36}$

   (b) $\frac{7}{12}$ (c) $\frac{14}{3}$, $\frac{10}{9}$

Miscellaneous Exercises

1. (a)

x	0	1	2
P	$\frac{25}{36}$	$\frac{10}{36}$	$\frac{1}{36}$

   (b)

x	1	2	3	4	5	6
P	$\frac{11}{36}$	$\frac{9}{36}$	$\frac{7}{36}$	$\frac{5}{36}$	$\frac{3}{36}$	$\frac{1}{36}$

   (c)

x	0	1	2	3
P	$\frac{1}{8}$	$\frac{3}{8}$	$\frac{3}{8}$	$\frac{1}{8}$

2. (a) 0.04 (b) 5 (c) 4

3. $P(X=x)=\frac{10-x}{45}$, $x=1, 2, \ldots, 9; 3\frac{2}{3}$, 2.21, 1

4. (a) $\frac{1}{36}$ (b) $\frac{5}{36}$ (c) $\frac{11}{36}$; $-\frac{1}{36}$, 7

5. $\frac{35}{18}$ (a) $\frac{1}{2}$ (b) $\frac{1}{12}$ (c) $\frac{1}{2}\left(\frac{1}{6}\right)^{r-1}$; $\frac{3}{5}$, £1.50

6. (a) $\frac{16}{81}$, $\frac{32}{81}$, $\frac{24}{81}$, $\frac{8}{81}$, $\frac{1}{81}$

   (b) −50p

7. $P(X=x)=\frac{1}{6}$, $x=$ 1, 2, 3, 4, 5

   $P(X=6)=0$

   $P(X=x)=\frac{1}{36}$, $x=7, 8, \ldots, 12$

   $4\frac{1}{12}$, $\frac{6}{17}$

8. (i) $\frac{1}{8}$, $\frac{5}{24}$ (ii) 2.78 (3 s.f.)

   (iii) 0.260 (3 s.f.)

5 BINOMIAL DISTRIBUTIONS

Exercise 5A

1. (a) 0.329 (b) 0.351 (c) 0.912
2. (a) 0.000138 (b) 0.0282 (c) 0.953
3. 1 $(P(R=1)=0.422)$
4. 0.230
5. 0.984

Activity 5

1. True
2. False (19th)
3. True
4. True
5. False (Van Gogh)
6. True

Exercise 5B

1. 0.263
2. 7
3. $\frac{2}{3}$
4. 0.51, 38
5. (a) $B(20, 0.2)$ (b) 4, 3.2 (c) 0.0026
7. B((18, 0.8), 15
8. (a) 0.000977 (b) 0.002197 ; 0.377
9. 0.3770
10. P (Don scores at least 3) = 0.3222
    P (Yvette scores at least 3) = 0.3174

Miscellaneous Exercises

1. (a) (i) 0.5289 (ii) 0.3158; 0.0301
   (b) No (c) Possibly

2. 0.1, 0.23 (2 d.p.)
3. (b) (i) 0.68 (ii) 8, 1.6
4. (a) 4.8, 0.98 (c) 0.737 (d) 0.388
5. (a) $\frac{2}{3}$ (b) 0.0424
6. (a) $(1-p)^8(36p^2+8p+1)$

   $(1-p)^5+5p(1-p)^8(1+4p)$

   (b) 0.678, 0.630, 0.0547, 0.0605

6 POISSON DISTRIBUTION

Exercise 6A

1. (a) 0.224 (b) 0.224 (c) 0.185
   (d) 0.423
2. $\lambda = 12$, 0.0127
3. 1.13, 0.0777
4. $X \sim Po(1)$
5. (a) 0.00674 (b) 0.0337 (c) 0.006746
   (d) 0.0337 (e) 0.125 (f) 0.125

Activity 4

$E(M)=3$, $V(M)=2.6$

$E(F)=2.48$, $V(F)=2.25$

$E(T)=5.48$, $V(T)=4.97$

Activity 7

(a) 0.149 (b) 0.815 (c) 0.313 (d) 0.018

Exercise 6B

1. (a) 0.2149 (b) 0.0424
   1.6
2. 0.8666; 0.9228; 0.04683
3. (a) 0.1755 (b) 0.5595 (c) 0.1024
   (d) 0.5343 (e) 0.00541; 10
4. (a) £14 (b) (i) 0.0743 (ii) 0.1931
   (iii) 0.2510 (iv) 0.4816
   (c) £10.70

Exercise 6C

1. Binomial (a) 0.368 (b) 0.368 (c) 0.0153
   Poisson (a) 0.368 (b) 0.368 (c) 0.0153
2. $X \sim Po(12)$ (a) 0.979 (b) 0.992
3. $Po(2)$ (a) 0.271 (b) 0.865
4. $Po(3)$, 0.950
5. (a) 0.190 (b) 0.826 (c) 0.884
6. 0.3712 (a) 0.2169 (b) 0.2646

Miscellaneous Exercises

1. (a) 0.713 (b) 0.433
2. 0.996
3. 3
4. (a) 4 (b) 0 (c) 13
5. (a) 0.826 (b) 0.407
6. 0.135 ; 1
7. (a) $6\frac{1}{3}$ (b) $1\frac{8}{9}$
8. 0.087
9. 0.4, 0.1, 0.41
10. 10
11. 0.010, 15
13. 0.050, 0.986
14. 0.181, 0.398
15. (a) 0.706 (b) $0.3027-0.1257 \approx 0.1772$.
16. Unlikely
17. (a) 0.387 (b) 0.929 (c) 0.893
    (d) 0.205 (e) 0.816 , 0.029
18. (a) 0.5152 (b) 0.0308 (c) 0.454
    (d) 0.333 (e) 0.265 (f) 0.4068

7 CONTINUOUS PROBABILITY DISTRIBUTIONS

Exercise 7A

1. (a) No (b) Yes (c) Yes (d) Yes (e) No
2. $A=\frac{3}{13}$

Exercise 7B

1. 0.375
2. 0.669
3. 0.040

Exercise 7C

1. $\frac{1}{2}$, $\frac{1}{20}$
2. $\frac{13}{6}$, $\frac{11}{36}$
3. $\frac{2}{5}$, $\frac{1}{25}$
4. $k=1$; 1, 1
5. $\frac{1}{108}$; $\frac{12}{5}$, $\frac{36}{25}$

Exercise 7D

1. (a) 2 (b) 1
2. (a) $4-2\sqrt{2}$ (b) ln2

Miscellaneous Exercises

1. $a=12$, $b=1$, $p=0.0523$
2. (a) $\frac{4}{27}$ (b) $\mu=\frac{9}{5}$, $s^2=\frac{9}{25}$ (c) 0.0272
3. (a) $\frac{3}{8}$ (c) $\frac{97}{60}$, $\frac{4}{3}$, median > mean
4. (b) 2, $\frac{1}{5}$ (c) $\frac{5}{32}$
5. (b) 6, 2 (c) $\frac{20}{3}$ (d) $\frac{56}{9}$ (e) 4.42
6. (a) $\frac{25}{36}$ (b) 0.6 hours (c) (i) $\frac{1}{8}$ (ii) 0.727
7. (a) 0.393 (b) 0.287 (c) 2.61
8. (b) $\frac{\pi}{6}$ (c) 0.64
9. (a) 200 hours (b) 0.4

8 THE NORMAL DISTRIBUTION

Exercise 8A

1. (a) 2% (b) 16% (c) 6

Exercise 8B

1. 0.20611
2. 0.79389
3. 0.79389
4. 0.20611
5. 0.58778
6. 0.68268
7. 0.83453
8. 0.49379
9. 0.02500
10. 0.95000

Exercise 8C

1. (a) 0.02275 (b) 0.08268 (c) 0.99865
2. (a) 0.15866 (b) 0.68268 (c) 0.62465
3. (a) 0.95254 (b) 0.25850 (c) 0.74927
4. 0.977
5. 0.1043
6. 0.57% ; 1
7. 0.067
8. 11.5%

Exercise 8D

1. 5%
2. 24
3. (a) 21/22 (b) about 10
4. 1.56, 2.7%
5. (a) 95.0, 3 (b) 0.99%
6. 398, 63

Exercise 8E

1. (a) 0.052 (b) 0.040
2. 0.0062
3. It is surprising
4. (a) 0.0276 (b) 0.008
5. 27

Miscellaneous Exercises

1. (a) 21.19% (b) 22.65 g
2. (a) 0.067 (b) 0.015 (c) 0.401
3. (a) 0.087 (b) 42
4. 174.18, 6.849, 0.098
5. (a) 0.2025 (b) 0.410 (c) 0.0238
6. (a) 0.115 (b) 33.1
7. (a) 0.68525 (b) 40.494 m (c) 40.06, 4.88
   (d) Gwen
8. (a) 0.82% (b) 95.59%
   0.052, A
9. (a) (i) 0.1912 (ii) 0.246
   (b) (i) 0.170 (ii) 0.648

9 ESTIMATION

Exercise 9A

1. Significant differences at 5% level (just).
2. Significant differences at 5% level.
3. No evidence to support contined growth.
4 Manufacturer is correct.

Activity 7

(a) (949, 1048) (b) 89% (c) 40

Miscellaneous Exercises

1. (a) 0.746 (b) (55.64, 58.56)
2. 502.4 ml
3. (217.9, 227.4), (211.5, 239.3), 31
4. $\bar{x}=5.7232$, $s=0.0397$, 0.00397,
   (5.7154, 5.7310), 379
5. (a) (i) (5.94, 9.56) (ii) 9.27
   (iii) (2.486, 13.014) (b) 18
6. (a) 391.3; about 251.7
   (b) [266.5, 516.1]

7. $(0.123, 0.392)$
   $(170.84, 178.16)$
   $(165.6, 186.8)$
   $[487.76, 530.59]$
8. $(1009.0, 1036.0)$
   $(1025.7, 1039.1)$
   $(£910.1, £940.9)$

10 HYPOTHESIS TESTING

Exercise 10A

1. $p(5, 6 \text{ or } 7) = 0.045$ - Yes, at 5%.
2. $p(2, 1 \text{ or } 0) = 0.167$ - Not sufficient evidence at 5%.
3. $p(7, 8, 9 \text{ or } 10) = 0.004$ - Yes, at 5%

Exercise 10B

1. $p(6 \text{ or more improvements}) = 0.1445$
   - not significant at 5%.
2. $p(9 \text{ or more improvements}) = 0.073$
   - not significant at 5%.
3. $p(7 \text{ or more improvements}) = 0.090$
   - not significant at 5%.
4. Not sufficient evidence to suggest that 4 star petrol is better.
5. $p(6 \text{ or more improvements}) = 0.0625$
   - not significant at 5%.

Exercise 10C

1. Justified; $(8.19, 8.53)$
2. $(45.6, 49.4)$, $\alpha > 4$, 0.0321

Miscellaneous Exercises

1. $z = -3.28$; critical $z = \pm 1.96$
   - significant difference.
2. $z = 1.56$; critical $z = 1.64$ - not sufficient evidence to suggest significant improvement.
3. $z = 2.61$; critical $z = 1.64$ - evidence suggests significant improvement.
4. $z = -3.10$; critical $z = 1.64$ - evidence to suggest that mean mass is less than 50 kg.
5. $z = -1.733$; critical $z = -1.64$ - reject H_0.
6. $z = -1.868$; critical $z = \pm 1.96$ - theory could be true.

11 CHI-SQUARED

Exercise 11A

1. $\chi^2 = 3.41$; critical $\chi^2 = 7.815$ - no evidence to suggest die is not fair.
2. $\chi^2 = 5.50$; critical $\chi^2 = 11.070$ - no evidence to suggest die is biased.
3. $\chi^2 = 10.94$; critical $\chi^2 = 9.488$ - differences significant.

Miscellaneous Exercises

1. $\chi^2 = 7.97$; critical $\chi^2 = 9.488$ - no evidence to suggest that patients attending the Centre depend on social class;

   $\chi^2 = 18.05$; critical $\chi^2 = 15.507$ (8 degrees of freedom), so there is evidence to suggest that the reason for visits to the surgery is not independent of social class.
2. (a) $\chi^2 = 23.79$; critical $\chi^2 = 9.210$ (2 degrees of freedom); evidence that proportion favouring expenditure on road improvements does depend on area;

   (b) Critical $\chi^2 = 5.99$ (2 degrees of freedom) so evidence to suggest that proportion favouring expenditure on road improvements is independent of area.
3. $\chi^2 = 17.01$; critical $\chi^2 = 5.991$ (2 degrees of freedom) so there is evidence to suggest that the proportion of transferred cases is not independent of the surgeon.
4. (a) $\chi^2 = 12.03$; critical $\chi^2 = 7.815$ (3 degrees of freedom) so evidence to suggest that the work is associated with the type of question.

   (b) Binomial and Poisson, since they have nearly equal responses.

   (c) $\chi^2 = 17.6$; critical $\chi^2 = 7.815$ (3 degrees of freedom) so evidence to suggest that the proportion of candidates attempting a question is associated with type of question.
5. (a) $\chi^2 = 14.77$; critical $\chi^2 = 13.277$ (4 degrees of freedom) so there is evidence to suggest that the number of books borrowed depends on the day of the week.

   (b) $\chi^2 = 16.93$; critical $\chi^2 = 13.277$ (4 degrees of freedom) so there is evidence to suggest that the length of employment is not independent of grade.

12 CORRELATION AND REGRESSION

Exercise 12A

1. (b) (i) $r=1$ (ii) $r=1$ (iii) $r=0$ (iv) $r=0$
2. (a) −0.676 (b) −0.676 (c) 0
3. $\bar{x}=56.9$, $\bar{y}=27.7$; (52,41), (53, 35), (61, 22);

   $s_x=34.404$, $s_y=15.966$, $r=0.880$
4. (i) 0.989
5. 0.995; (25, 106.1); 1

Exercise 12B

1. (i) 0.814 (ii) 0.792
2. −0.486, $P(\Sigma d^2 \geq 52)=0.178 \Rightarrow$ not significant
3. $0.5\dot{2}\dot{7}$; $P(\Sigma d^2 \leq 78)=0.0616$; No
4. 0.5
5. (a) (i) 0.976

   (ii) 0.304

Exercise 12C

1. $y=7.471x-7.714$

   (a) 478 (b) 739 (c) 7464

   last value is outside range of data.
2. $y=3.179x+4.221$; 3.179 kg
3. $y=0.089x+0.755$; 4.5 kg; −0.1 kg impossible.

Exercise 12D

1. $y=1.03x+0.53$
2. (b) $y=4.124\theta+23.334$ (c) (10.5, 91)

   (d) (i) 72.8 cm (ii) 188.3 cm
3. $H=1.579T-11.028$

   $H=1.3$
4. $y=0.451x-0.660$

   $x=1.306y+2.145$

   $y=2$, $x=3$

Miscellaneous Exercises

1. (b) $y=-35.7+0.953x$

   (c) (i) 131 (ii) 141 (iii) 250
2. (a) 0.460 (b) 0.75
3. (a) −0.589 (c) 0.923
4. (b) $y=4.48x+209$
5. (a) 0.312 (b) $r=0.469$
6. $y=1.17x+127.1$
7. (b) 0.296 (c) E and G; 0.999
8. (b) $y=0.389x-0.142$

   (c) 23.2°C

   (d) (i) $y=23.4t-0.142$

   (ii) $z=272.9+0.389x$
9. (a) $y=0.584x+44.3$, $x=0.949y+37.4$

   (c) $(\bar{x}, \bar{y})$

   (d) 152.3, 184.5

   (e) data points lie on a straight line
10. (b) $y=3713+191.81x$

    (d) (i) $a=4213$

    (ii) $a=4010$, $b=207.15$

    (iii) $a=4162$, $b=199.48$
11. (a) 0.745
12. (c) −0.824

APPENDIX

TABLE 1 Random Digits

TABLE 2 Cumulative Binomial Distribution Functions

TABLE 3 Cumulative Poisson Distribution Functions

TABLE 4 Normal Distribution Function

TABLE 5 Percentage Points of the Normal Distribution

TABLE 6 χ^2 Distribution Function

TABLE 7 Critical Values of the Product Moment Correlation Coefficient

TABLE 8 Critical Values of Spearman's Rank Correlation Coefficient

Sources of Data

Further Reading

TABLE 1 Random Digits

38956	29927	66187	80784	37542	62446	13481	72730	48511	42315
94451	62506	22780	30720	79338	68358	62765	33401	82758	42929
01323	83752	10664	12193	88766	76763	90977	46881	59089	39648
81916	75703	27522	79504	06662	25468	92407	19626	61173	52793
83727	99617	59120	33554	32904	95312	61763	68868	94179	73442
51966	17490	64900	12690	95474	53849	64791	35843	44832	01296
27355	02384	16680	76637	42437	27994	24718	09566	43821	89315
78802	44031	51668	85907	22683	06119	25360	35480	91334	01522
46134	94058	36466	99717	57651	02512	98785	86491	76812	10324
57217	88783	77127	95783	40666	82539	84224	94354	41979	32823
67895	33380	47444	02936	57303	31458	28669	22538	66884	38370
99108	95198	41684	89066	17963	39042	50791	44683	15134	19909
79310	03183	62706	65531	47767	42347	51899	33582	28098	43168
34447	26623	00550	52329	90292	37508	97310	92049	47365	80242
02737	57929	14290	08118	95473	91586	58953	74998	73950	54662
15269	49103	92150	78211	27762	18135	43479	61698	77768	00223
96198	98634	31870	56839	60478	62129	87149	60240	09079	38567
75823	90593	76248	60379	98204	59254	51616	41091	11818	11001
19611	68604	90298	38595	52048	95137	73363	53307	37914	27903
32205	72711	43441	87108	82155	43650	81967	56348	19878	75813
74513	08193	05302	11352	48369	55731	81158	21037	29534	98074
39851	74829	51695	51682	97660	97110	69540	69776	22736	54635
10349	25900	81265	25339	43875	38563	43530	36289	78810	18959
39871	42417	50106	24752	94664	11611	05720	77091	96338	68507
51268	32291	57653	42135	36440	79427	11660	15666	55682	25449
07468	24096	57419	35611	91179	51464	94284	92449	97347	22184
34454	50344	22824	09193	98771	30963	02876	97671	56397	91677
41503	76672	52872	48610	31314	21545	23601	18278	93530	02142
47261	50385	70112	26897	00077	04803	98326	88933	17710	75750
14852	64222	95920	80534	55090	04105	01415	11376	20709	78887
51198	11602	06891	07924	42959	73124	36830	70559	55739	73191
20818	87962	92071	13405	05057	85947	73043	94208	52829	88272
77297	41595	07611	36646	70863	57797	82033	19236	74608	14324
64648	34917	58038	47230	38817	70605	62771	02851	23195	20204
49898	50622	76133	54065	34055	13961	07604	30260	92240	40736
95060	14422	58282	73673	04535	03557	40036	85475	16021	77173
74300	48254	71043	44942	12252	59557	53013	26170	21980	18582
62710	59322	65251	84379	05985	45765	38349	68661	18129	29338
01352	04224	19593	72554	54239	44870	38726	51297	82412	65799
95076	17264	41154	16019	70481	97716	53185	53901	89036	01253
62445	09632	07182	78111	19253	12414	73496	24090	54974	48941
86267	54282	74626	40866	91371	44589	31478	58842	71961	38487
69681	80207	43497	37079	53974	20241	62576	15660	68405	57982
51884	93899	94309	56732	59858	28457	74546	45424	92496	71035
80038	46869	52284	00000	42554	58770	83458	58425	60956	21595
25342	61693	10160	27212	91407	61420	55196	32064	99083	45348
87696	88047	21252	52766	88011	96661	77691	78801	05384	92340
25749	27087	84246	04208	37579	54270	94698	86310	06727	88176
15251	34691	89127	51214	38276	27601	02422	77625	02017	13801
64230	48467	55548	84036	63668	20271	26235	76671	51372	35552

TABLE 2 Cumulative Binomial Distribution Functions

These tables give the probability of obtaining at most x successes in n independent trials, when the probability of success for each trial is p.

x \ p	0.01	0.02	0.03	0.04	0.05	0.06	0.07	0.08	0.09	0.10	0.15	0.20	0.25	0.30	0.35	0.40	0.45	0.50
$n = 5$																		
0	0.9510	0.9039	0.8587	0.8154	0.7738	0.7339	0.6957	0.6591	0.6240	0.5905	0.4437	0.3277	0.2373	0.1681	0.1160	0.0778	0.0503	0.0313
1	0.9990	0.9962	0.9915	0.9852	0.9774	0.9681	0.9575	0.9456	0.9326	0.9185	0.8352	0.7373	0.6328	0.5282	0.4284	0.3370	0.2562	0.1875
2	1.0000	0.9999	0.9997	0.9994	0.9988	0.9980	0.9969	0.9955	0.9937	0.9914	0.9734	0.9421	0.8965	0.8369	0.7648	0.6826	0.5931	0.5000
3	1.0000	1.0000	1.0000	1.0000	1.0000	0.9999	0.9999	0.9998	0.9997	0.9995	0.9978	0.9933	0.9844	0.9692	0.9460	0.9130	0.8688	0.8125
4	1.0000	1.0000	1.0000	1.0000	1.0000	1.0000	1.0000	1.0000	1.0000	1.0000	0.9999	0.9997	0.9990	0.9976	0.9947	0.9898	0.9815	0.9688
5	1.0000	1.0000	1.0000	1.0000	1.0000	1.0000	1.0000	1.0000	1.0000	1.0000	1.0000	1.0000	1.0000	1.0000	1.0000	1.0000	1.0000	1.0000
$n = 6$																		
0	0.9415	0.8858	0.8330	0.7828	0.7351	0.6899	0.6470	0.6064	0.5679	0.5314	0.3771	0.2621	0.1780	0.1176	0.0754	0.0467	0.0277	0.0156
1	0.9985	0.9943	0.9875	0.9784	0.9672	0.9541	0.9392	0.9227	0.9048	0.8857	0.7765	0.6554	0.5339	0.4202	0.3191	0.2333	0.1636	0.1094
2	1.0000	0.9998	0.9995	0.9988	0.9978	0.9962	0.9942	0.9915	0.9882	0.9842	0.9527	0.9011	0.8306	0.7443	0.6471	0.5443	0.4415	0.3438
3	1.0000	1.0000	1.0000	1.0000	0.9999	0.9998	0.9997	0.9995	0.9992	0.9987	0.9941	0.9830	0.9624	0.9295	0.8826	0.8208	0.7447	0.6563
4	1.0000	1.0000	1.0000	1.0000	1.0000	1.0000	1.0000	1.0000	1.0000	0.9999	0.9996	0.9984	0.9954	0.9891	0.9777	0.9590	0.9308	0.8906
5	1.0000	1.0000	1.0000	1.0000	1.0000	1.0000	1.0000	1.0000	1.0000	1.0000	1.0000	0.9999	0.9998	0.9993	0.9982	0.9959	0.9917	0.9844
6	1.0000	1.0000	1.0000	1.0000	1.0000	1.0000	1.0000	1.0000	1.0000	1.0000	1.0000	1.0000	1.0000	1.0000	1.0000	1.0000	1.0000	1.0000
$n = 7$																		
0	0.9321	0.8681	0.8080	0.7514	0.6983	0.6485	0.6017	0.5578	0.5168	0.4783	0.3206	0.2097	0.1335	0.0824	0.0490	0.0280	0.0152	0.0078
1	0.9980	0.9921	0.9829	0.9706	0.9556	0.9382	0.9187	0.8974	0.8745	0.8503	0.7166	0.5767	0.4449	0.3294	0.2338	0.1586	0.1024	0.0625
2	1.0000	0.9997	0.9991	0.9980	0.9962	0.9937	0.9903	0.9860	0.9807	0.9743	0.9262	0.8520	0.7564	0.6471	0.5323	0.4199	0.3164	0.2266
3	1.0000	1.0000	1.0000	0.9999	0.9998	0.9996	0.9993	0.9988	0.9982	0.9973	0.9879	0.9667	0.9294	0.8740	0.8002	0.7102	0.6083	0.5000
4	1.0000	1.0000	1.0000	1.0000	1.0000	1.0000	1.0000	0.9999	0.9999	0.9998	0.9988	0.9953	0.9871	0.9712	0.9444	0.9037	0.8471	0.7734
5	1.0000	1.0000	1.0000	1.0000	1.0000	1.0000	1.0000	1.0000	1.0000	1.0000	0.9999	0.9996	0.9987	0.9962	0.9910	0.9812	0.9643	0.9375
6	1.0000	1.0000	1.0000	1.0000	1.0000	1.0000	1.0000	1.0000	1.0000	1.0000	1.0000	1.0000	0.9999	0.9998	0.9994	0.9984	0.9963	0.9922
7	1.0000	1.0000	1.0000	1.0000	1.0000	1.0000	1.0000	1.0000	1.0000	1.0000	1.0000	1.0000	1.0000	1.0000	1.0000	1.0000	1.0000	1.0000
$n = 8$																		
0	0.9227	0.8508	0.7837	0.7214	0.6634	0.6096	0.5596	0.5132	0.4703	0.4305	0.2725	0.1678	0.1001	0.0576	0.0319	0.0168	0.0084	0.0039
1	0.9973	0.9897	0.9777	0.9619	0.9428	0.9208	0.8965	0.8702	0.8423	0.8131	0.6572	0.5033	0.3671	0.2553	0.1691	0.1064	0.0632	0.0352
2	0.9999	0.9996	0.9987	0.9969	0.9942	0.9904	0.9853	0.9789	0.9711	0.9619	0.8948	0.7969	0.6785	0.5518	0.4278	0.3154	0.2201	0.1445
3	1.0000	1.0000	0.9999	0.9998	0.9996	0.9993	0.9987	0.9978	0.9966	0.9950	0.9786	0.9437	0.8862	0.8059	0.7064	0.5941	0.4770	0.3633
4	1.0000	1.0000	1.0000	1.0000	1.0000	1.0000	0.9999	0.9999	0.9997	0.9996	0.9971	0.9896	0.9727	0.9420	0.8939	0.8263	0.7396	0.6367
5	1.0000	1.0000	1.0000	1.0000	1.0000	1.0000	1.0000	1.0000	1.0000	1.0000	0.9998	0.9988	0.9958	0.9887	0.9747	0.9502	0.9115	0.8555
6	1.0000	1.0000	1.0000	1.0000	1.0000	1.0000	1.0000	1.0000	1.0000	1.0000	1.0000	0.9999	0.9996	0.9987	0.9964	0.9915	0.9819	0.9648
7	1.0000	1.0000	1.0000	1.0000	1.0000	1.0000	1.0000	1.0000	1.0000	1.0000	1.0000	1.0000	1.0000	0.9999	0.9998	0.9993	0.9983	0.9961
8	1.0000	1.0000	1.0000	1.0000	1.0000	1.0000	1.0000	1.0000	1.0000	1.0000	1.0000	1.0000	1.0000	1.0000	1.0000	1.0000	1.0000	1.0000

p / x	0.01	0.02	0.03	0.04	0.05	0.06	0.07	0.08	0.09	0.10	0.15	0.20	0.25	0.30	0.35	0.40	0.45	0.50
$n = 9$																		
0	0.9135	0.8337	0.7602	0.6925	0.6302	0.5730	0.5204	0.4722	0.4279	0.3874	0.2316	0.1342	0.0751	0.0404	0.0207	0.0101	0.0046	0.0020
1	0.9966	0.9869	0.9718	0.9522	0.9288	0.9022	0.8729	0.8417	0.8088	0.7748	0.5995	0.4362	0.3003	0.1960	0.1211	0.0705	0.0385	0.0195
2	0.9999	0.9994	0.9980	0.9955	0.9916	0.9862	0.9791	0.9702	0.9595	0.9470	0.8591	0.7382	0.6007	0.4628	0.3373	0.2318	0.1495	0.0898
3	1.0000	1.0000	0.9999	0.9997	0.9994	0.9987	0.9977	0.9963	0.9943	0.9917	0.9661	0.9144	0.8343	0.7297	0.6089	0.4826	0.3614	0.2539
4	1.0000	1.0000	1.0000	1.0000	1.0000	0.9999	0.9998	0.9997	0.9995	0.9991	0.9944	0.9804	0.9511	0.9012	0.8283	0.7334	0.6214	0.5000
5	1.0000	1.0000	1.0000	1.0000	1.0000	1.0000	1.0000	1.0000	1.0000	0.9999	0.9994	0.9969	0.9900	0.9747	0.9464	0.9006	0.8342	0.7461
6	1.0000	1.0000	1.0000	1.0000	1.0000	1.0000	1.0000	1.0000	1.0000	1.0000	1.0000	0.9997	0.9987	0.9957	0.9888	0.9750	0.9502	0.9102
7	1.0000	1.0000	1.0000	1.0000	1.0000	1.0000	1.0000	1.0000	1.0000	1.0000	1.0000	1.0000	0.9999	0.9996	0.9986	0.9962	0.9909	0.9805
8	1.0000	1.0000	1.0000	1.0000	1.0000	1.0000	1.0000	1.0000	1.0000	1.0000	1.0000	1.0000	1.0000	1.0000	0.9999	0.9997	0.9992	0.9980
9	1.0000	1.0000	1.0000	1.0000	1.0000	1.0000	1.0000	1.0000	1.0000	1.0000	1.0000	1.0000	1.0000	1.0000	1.0000	1.0000	1.0000	1.0000
$n = 10$																		
0	0.9044	0.8171	0.7374	0.6648	0.5987	0.5386	0.4840	0.4344	0.3894	0.3487	0.1969	0.1074	0.0563	0.0282	0.0135	0.0060	0.0025	0.0010
1	0.9957	0.9838	0.9655	0.9418	0.9139	0.8824	0.8483	0.8121	0.7746	0.7361	0.5443	0.3758	0.2440	0.1493	0.0860	0.0464	0.0233	0.0107
2	0.9999	0.9991	0.9972	0.9938	0.9885	0.9812	0.9717	0.9599	0.9460	0.9298	0.8202	0.6778	0.5256	0.3828	0.2616	0.1673	0.0996	0.0547
3	1.0000	1.0000	0.9999	0.9996	0.9990	0.9980	0.9964	0.9942	0.9912	0.9872	0.9500	0.8791	0.7759	0.6496	0.5138	0.3823	0.2660	0.1719
4	1.0000	1.0000	1.0000	1.0000	0.9999	0.9998	0.9997	0.9994	0.9990	0.9984	0.9901	0.9672	0.9219	0.8497	0.7515	0.6331	0.5044	0.3770
5	1.0000	1.0000	1.0000	1.0000	1.0000	1.0000	1.0000	1.0000	0.9999	0.9999	0.9986	0.9936	0.9803	0.9527	0.9051	0.8338	0.7384	0.6230
6	1.0000	1.0000	1.0000	1.0000	1.0000	1.0000	1.0000	1.0000	1.0000	1.0000	0.9999	0.9991	0.9965	0.9894	0.9740	0.9452	0.8980	0.8281
7	1.0000	1.0000	1.0000	1.0000	1.0000	1.0000	1.0000	1.0000	1.0000	1.0000	1.0000	0.9999	0.9996	0.9984	0.9952	0.9877	0.9726	0.9453
8	1.0000	1.0000	1.0000	1.0000	1.0000	1.0000	1.0000	1.0000	1.0000	1.0000	1.0000	1.0000	1.0000	0.9999	0.9995	0.9983	0.9955	0.9893
9	1.0000	1.0000	1.0000	1.0000	1.0000	1.0000	1.0000	1.0000	1.0000	1.0000	1.0000	1.0000	1.0000	1.0000	1.0000	0.9999	0.9997	0.9990
10	1.0000	1.0000	1.0000	1.0000	1.0000	1.0000	1.0000	1.0000	1.0000	1.0000	1.0000	1.0000	1.0000	1.0000	1.0000	1.0000	1.0000	1.0000
$n = 20$																		
0	0.8179	0.6676	0.5438	0.4420	0.3585	0.2901	0.2342	0.1887	0.1516	0.1216	0.0388	0.0115	0.0032	0.0008	0.0002	0.0000	0.0000	0.0000
1	0.9831	0.9401	0.8802	0.8103	0.7358	0.6605	0.5869	0.5169	0.4516	0.3917	0.1756	0.0692	0.0243	0.0076	0.0021	0.0005	0.0001	0.0000
2	0.9990	0.9929	0.9790	0.9561	0.9245	0.8850	0.8390	0.7879	0.7334	0.6769	0.4049	0.2061	0.0913	0.0355	0.0121	0.0036	0.0009	0.0002
3	1.0000	0.9994	0.9973	0.9926	0.9841	0.9710	0.9529	0.9294	0.9007	0.8670	0.6477	0.4114	0.2252	0.1071	0.0444	0.0160	0.0049	0.0013
4	1.0000	1.0000	0.9997	0.9990	0.9974	0.9944	0.9893	0.9817	0.9710	0.9568	0.8298	0.6296	0.4148	0.2375	0.1182	0.0510	0.0189	0.0059
5	1.0000	1.0000	1.0000	0.9999	0.9997	0.9991	0.9981	0.9962	0.9932	0.9887	0.9327	0.8042	0.6172	0.4164	0.2454	0.1256	0.0553	0.0207
6	1.0000	1.0000	1.0000	1.0000	1.0000	0.9999	0.9997	0.9994	0.9987	0.9976	0.9781	0.9133	0.7858	0.6080	0.4166	0.2500	0.1299	0.0577
7	1.0000	1.0000	1.0000	1.0000	1.0000	1.0000	1.0000	0.9999	0.9998	0.9996	0.9941	0.9679	0.8982	0.7723	0.6010	0.4159	0.2520	0.1316
8	1.0000	1.0000	1.0000	1.0000	1.0000	1.0000	1.0000	1.0000	1.0000	0.9999	0.9987	0.9900	0.9591	0.8867	0.7624	0.5956	0.4143	0.2517
9	1.0000	1.0000	1.0000	1.0000	1.0000	1.0000	1.0000	1.0000	1.0000	1.0000	0.9998	0.9974	0.9861	0.9520	0.8782	0.7553	0.5914	0.4119
10	1.0000	1.0000	1.0000	1.0000	1.0000	1.0000	1.0000	1.0000	1.0000	1.0000	1.0000	0.9994	0.9961	0.9829	0.9468	0.8725	0.7507	0.5881
11	1.0000	1.0000	1.0000	1.0000	1.0000	1.0000	1.0000	1.0000	1.0000	1.0000	1.0000	0.9999	0.9991	0.9949	0.9804	0.9435	0.8692	0.7483
12	1.0000	1.0000	1.0000	1.0000	1.0000	1.0000	1.0000	1.0000	1.0000	1.0000	1.0000	1.0000	0.9998	0.9987	0.9940	0.9790	0.9420	0.8684
13	1.0000	1.0000	1.0000	1.0000	1.0000	1.0000	1.0000	1.0000	1.0000	1.0000	1.0000	1.0000	1.0000	0.9997	0.9985	0.9935	0.9786	0.9423
14	1.0000	1.0000	1.0000	1.0000	1.0000	1.0000	1.0000	1.0000	1.0000	1.0000	1.0000	1.0000	1.0000	1.0000	0.9997	0.9984	0.9936	0.9793
15	1.0000	1.0000	1.0000	1.0000	1.0000	1.0000	1.0000	1.0000	1.0000	1.0000	1.0000	1.0000	1.0000	1.0000	1.0000	0.9997	0.9985	0.9941
16	1.0000	1.0000	1.0000	1.0000	1.0000	1.0000	1.0000	1.0000	1.0000	1.0000	1.0000	1.0000	1.0000	1.0000	1.0000	1.0000	0.9997	0.9987
17	1.0000	1.0000	1.0000	1.0000	1.0000	1.0000	1.0000	1.0000	1.0000	1.0000	1.0000	1.0000	1.0000	1.0000	1.0000	1.0000	1.0000	0.9998
18	1.0000	1.0000	1.0000	1.0000	1.0000	1.0000	1.0000	1.0000	1.0000	1.0000	1.0000	1.0000	1.0000	1.0000	1.0000	1.0000	1.0000	1.0000
19	1.0000	1.0000	1.0000	1.0000	1.0000	1.0000	1.0000	1.0000	1.0000	1.0000	1.0000	1.0000	1.0000	1.0000	1.0000	1.0000	1.0000	1.0000
20	1.0000	1.0000	1.0000	1.0000	1.0000	1.0000	1.0000	1.0000	1.0000	1.0000	1.0000	1.0000	1.0000	1.0000	1.0000	1.0000	1.0000	1.0000

p	0.01	0.02	0.03	0.04	0.05	0.06	0.07	0.08	0.09	0.10	0.15	0.20	0.25	0.30	0.35	0.40	0.45	0.50	p
x	$n = 30$																		x
0	0.7397	0.5455	0.4010	0.2939	0.2146	0.1563	0.1134	0.0820	0.0591	0.0424	0.0076	0.0012	0.0002	0.0000	0.0000	0.0000	0.0000	0.0000	**0**
1	0.9639	0.8795	0.7731	0.6612	0.5535	0.4555	0.3694	0.2958	0.2343	0.1837	0.0480	0.0105	0.0020	0.0003	0.0000	0.0000	0.0000	0.0000	**1**
2	0.9967	0.9783	0.9399	0.8831	0.8122	0.7324	0.6487	0.5654	0.4855	0.4114	0.1514	0.0442	0.0106	0.0021	0.0003	0.0000	0.0000	0.0000	**2**
3	0.9998	0.9971	0.9881	0.9694	0.9392	0.8974	0.8450	0.7842	0.7175	0.6474	0.3217	0.1227	0.0374	0.0093	0.0019	0.0003	0.0000	0.0000	**3**
4	1.0000	0.9997	0.9982	0.9937	0.9844	0.9685	0.9447	0.9126	0.8723	0.8245	0.5245	0.2552	0.0979	0.0302	0.0075	0.0015	0.0002	0.0000	**4**
5	1.0000	1.0000	0.9998	0.9989	0.9967	0.9921	0.9838	0.9707	0.9519	0.9268	0.7106	0.4275	0.2026	0.0766	0.0233	0.0057	0.0011	0.0002	**5**
6	1.0000	1.0000	1.0000	0.9999	0.9994	0.9983	0.9960	0.9918	0.9848	0.9742	0.8474	0.6070	0.3481	0.1595	0.0586	0.0172	0.0040	0.0007	**6**
7	1.0000	1.0000	1.0000	1.0000	0.9999	0.9997	0.9992	0.9980	0.9959	0.9922	0.9302	0.7608	0.5143	0.2814	0.1238	0.0435	0.0121	0.0026	**7**
8	1.0000	1.0000	1.0000	1.0000	1.0000	1.0000	0.9999	0.9996	0.9990	0.9980	0.9722	0.8713	0.6736	0.4315	0.2247	0.0940	0.0312	0.0081	**8**
9	1.0000	1.0000	1.0000	1.0000	1.0000	1.0000	1.0000	0.9999	0.9998	0.9995	0.9903	0.9389	0.8034	0.5888	0.3575	0.1763	0.0694	0.0214	**9**
10	1.0000	1.0000	1.0000	1.0000	1.0000	1.0000	1.0000	1.0000	1.0000	0.9999	0.9971	0.9744	0.8943	0.7304	0.5078	0.2915	0.1350	0.0494	**10**
11	1.0000	1.0000	1.0000	1.0000	1.0000	1.0000	1.0000	1.0000	1.0000	1.0000	0.9992	0.9905	0.9493	0.8407	0.6548	0.4311	0.2327	0.1002	**11**
12	1.0000	1.0000	1.0000	1.0000	1.0000	1.0000	1.0000	1.0000	1.0000	1.0000	0.9998	0.9969	0.9784	0.9155	0.7802	0.5785	0.3592	0.1808	**12**
13	1.0000	1.0000	1.0000	1.0000	1.0000	1.0000	1.0000	1.0000	1.0000	1.0000	1.0000	0.9991	0.9918	0.9599	0.8737	0.7145	0.5025	0.2923	**13**
14	1.0000	1.0000	1.0000	1.0000	1.0000	1.0000	1.0000	1.0000	1.0000	1.0000	1.0000	0.9998	0.9973	0.9831	0.9348	0.8246	0.6448	0.4278	**14**
15	1.0000	1.0000	1.0000	1.0000	1.0000	1.0000	1.0000	1.0000	1.0000	1.0000	1.0000	0.9999	0.9992	0.9936	0.9699	0.9029	0.7691	0.5722	**15**
16	1.0000	1.0000	1.0000	1.0000	1.0000	1.0000	1.0000	1.0000	1.0000	1.0000	1.0000	1.0000	0.9998	0.9979	0.9876	0.9519	0.8644	0.7077	**16**
17	1.0000	1.0000	1.0000	1.0000	1.0000	1.0000	1.0000	1.0000	1.0000	1.0000	1.0000	1.0000	0.9999	0.9994	0.9955	0.9788	0.9286	0.8192	**17**
18	1.0000	1.0000	1.0000	1.0000	1.0000	1.0000	1.0000	1.0000	1.0000	1.0000	1.0000	1.0000	1.0000	0.9998	0.9986	0.9917	0.9666	0.8998	**18**
19	1.0000	1.0000	1.0000	1.0000	1.0000	1.0000	1.0000	1.0000	1.0000	1.0000	1.0000	1.0000	1.0000	1.0000	0.9996	0.9971	0.9862	0.9506	**19**
20	1.0000	1.0000	1.0000	1.0000	1.0000	1.0000	1.0000	1.0000	1.0000	1.0000	1.0000	1.0000	1.0000	1.0000	0.9999	0.9991	0.9950	0.9786	**20**
21	1.0000	1.0000	1.0000	1.0000	1.0000	1.0000	1.0000	1.0000	1.0000	1.0000	1.0000	1.0000	1.0000	1.0000	1.0000	0.9998	0.9984	0.9919	**21**
22	1.0000	1.0000	1.0000	1.0000	1.0000	1.0000	1.0000	1.0000	1.0000	1.0000	1.0000	1.0000	1.0000	1.0000	1.0000	1.0000	0.9996	0.9974	**22**
23	1.0000	1.0000	1.0000	1.0000	1.0000	1.0000	1.0000	1.0000	1.0000	1.0000	1.0000	1.0000	1.0000	1.0000	1.0000	1.0000	0.9999	0.9993	**23**
24	1.0000	1.0000	1.0000	1.0000	1.0000	1.0000	1.0000	1.0000	1.0000	1.0000	1.0000	1.0000	1.0000	1.0000	1.0000	1.0000	1.0000	0.9998	**24**
25	1.0000	1.0000	1.0000	1.0000	1.0000	1.0000	1.0000	1.0000	1.0000	1.0000	1.0000	1.0000	1.0000	1.0000	1.0000	1.0000	1.0000	1.0000	**25**

p	0.01	0.02	0.03	0.04	0.05	0.06	0.07	0.08	0.09	0.10	0.15	0.20	0.25	0.30	0.35	0.40	0.45	0.50	p
x	$n = 40$																		x
0	0.6690	0.4457	0.2957	0.1954	0.1285	0.0842	0.0549	0.0356	0.0230	0.0148	0.0015	0.0001	0.0000	0.0000	0.0000	0.0000	0.0000	0.0000	0
1	0.9393	0.8095	0.6615	0.5210	0.3991	0.2990	0.2201	0.1594	0.1140	0.0805	0.0121	0.0015	0.0001	0.0000	0.0000	0.0000	0.0000	0.0000	1
2	0.9925	0.9543	0.8822	0.7855	0.6767	0.5665	0.4625	0.3694	0.2894	0.2228	0.0486	0.0079	0.0010	0.0001	0.0000	0.0000	0.0000	0.0000	2
3	0.9993	0.9918	0.9686	0.9252	0.8619	0.7827	0.6937	0.6007	0.5092	0.4231	0.1302	0.0285	0.0047	0.0006	0.0001	0.0000	0.0000	0.0000	3
4	1.0000	0.9988	0.9933	0.9790	0.9520	0.9104	0.8546	0.7868	0.7103	0.6290	0.2633	0.0759	0.0160	0.0026	0.0003	0.0000	0.0000	0.0000	4
5	1.0000	0.9999	0.9988	0.9951	0.9861	0.9691	0.9419	0.9033	0.8535	0.7937	0.4325	0.1613	0.0433	0.0086	0.0013	0.0001	0.0000	0.0000	5
6	1.0000	1.0000	0.9998	0.9990	0.9966	0.9909	0.9801	0.9624	0.9361	0.9005	0.6067	0.2859	0.0962	0.0238	0.0044	0.0006	0.0001	0.0000	6
7	1.0000	1.0000	1.0000	0.9998	0.9993	0.9977	0.9942	0.9873	0.9758	0.9581	0.7559	0.4371	0.1820	0.0553	0.0124	0.0021	0.0002	0.0000	7
8	1.0000	1.0000	1.0000	1.0000	0.9999	0.9995	0.9985	0.9963	0.9919	0.9845	0.8646	0.5931	0.2998	0.1110	0.0303	0.0061	0.0009	0.0001	8
9	1.0000	1.0000	1.0000	1.0000	1.0000	0.9999	0.9997	0.9990	0.9976	0.9949	0.9328	0.7318	0.4395	0.1959	0.0644	0.0156	0.0027	0.0003	9
10	1.0000	1.0000	1.0000	1.0000	1.0000	1.0000	0.9999	0.9998	0.9994	0.9985	0.9701	0.8392	0.5839	0.3087	0.1215	0.0352	0.0074	0.0011	10
11	1.0000	1.0000	1.0000	1.0000	1.0000	1.0000	1.0000	1.0000	0.9999	0.9996	0.9880	0.9125	0.7151	0.4406	0.2053	0.0709	0.0179	0.0032	11
12	1.0000	1.0000	1.0000	1.0000	1.0000	1.0000	1.0000	1.0000	1.0000	0.9999	0.9957	0.9568	0.8209	0.5772	0.3143	0.1285	0.0386	0.0083	12
13	1.0000	1.0000	1.0000	1.0000	1.0000	1.0000	1.0000	1.0000	1.0000	1.0000	0.9986	0.9806	0.8968	0.7032	0.4408	0.2112	0.0751	0.0192	13
14	1.0000	1.0000	1.0000	1.0000	1.0000	1.0000	1.0000	1.0000	1.0000	1.0000	0.9996	0.9921	0.9456	0.8074	0.5721	0.3174	0.1326	0.0403	14
15	1.0000	1.0000	1.0000	1.0000	1.0000	1.0000	1.0000	1.0000	1.0000	1.0000	0.9999	0.9971	0.9738	0.8849	0.6946	0.4402	0.2142	0.0769	15
16	1.0000	1.0000	1.0000	1.0000	1.0000	1.0000	1.0000	1.0000	1.0000	1.0000	1.0000	0.9990	0.9884	0.9367	0.7978	0.5681	0.3185	0.1341	16
17	1.0000	1.0000	1.0000	1.0000	1.0000	1.0000	1.0000	1.0000	1.0000	1.0000	1.0000	0.9997	0.9953	0.9680	0.8761	0.6885	0.4391	0.2148	17
18	1.0000	1.0000	1.0000	1.0000	1.0000	1.0000	1.0000	1.0000	1.0000	1.0000	1.0000	0.9999	0.9983	0.9852	0.9301	0.7911	0.5651	0.3179	18
19	1.0000	1.0000	1.0000	1.0000	1.0000	1.0000	1.0000	1.0000	1.0000	1.0000	1.0000	1.0000	0.9994	0.9937	0.9637	0.8702	0.6844	0.4373	19
20	1.0000	1.0000	1.0000	1.0000	1.0000	1.0000	1.0000	1.0000	1.0000	1.0000	1.0000	1.0000	0.9998	0.9976	0.9827	0.9256	0.7870	0.5627	20
21	1.0000	1.0000	1.0000	1.0000	1.0000	1.0000	1.0000	1.0000	1.0000	1.0000	1.0000	1.0000	1.0000	0.9991	0.9925	0.9608	0.8669	0.6821	21
22	1.0000	1.0000	1.0000	1.0000	1.0000	1.0000	1.0000	1.0000	1.0000	1.0000	1.0000	1.0000	1.0000	0.9997	0.9970	0.9811	0.9233	0.7852	22
23	1.0000	1.0000	1.0000	1.0000	1.0000	1.0000	1.0000	1.0000	1.0000	1.0000	1.0000	1.0000	1.0000	0.9999	0.9989	0.9917	0.9595	0.8659	23
24	1.0000	1.0000	1.0000	1.0000	1.0000	1.0000	1.0000	1.0000	1.0000	1.0000	1.0000	1.0000	1.0000	1.0000	0.9996	0.9966	0.9804	0.9231	24
25	1.0000	1.0000	1.0000	1.0000	1.0000	1.0000	1.0000	1.0000	1.0000	1.0000	1.0000	1.0000	1.0000	1.0000	0.9999	0.9988	0.9914	0.9597	25
26	1.0000	1.0000	1.0000	1.0000	1.0000	1.0000	1.0000	1.0000	1.0000	1.0000	1.0000	1.0000	1.0000	1.0000	1.0000	0.9996	0.9966	0.9808	26
27	1.0000	1.0000	1.0000	1.0000	1.0000	1.0000	1.0000	1.0000	1.0000	1.0000	1.0000	1.0000	1.0000	1.0000	1.0000	0.9999	0.9988	0.9917	27
28	1.0000	1.0000	1.0000	1.0000	1.0000	1.0000	1.0000	1.0000	1.0000	1.0000	1.0000	1.0000	1.0000	1.0000	1.0000	1.0000	0.9996	0.9968	28
29	1.0000	1.0000	1.0000	1.0000	1.0000	1.0000	1.0000	1.0000	1.0000	1.0000	1.0000	1.0000	1.0000	1.0000	1.0000	1.0000	0.9999	0.9989	29
30	1.0000	1.0000	1.0000	1.0000	1.0000	1.0000	1.0000	1.0000	1.0000	1.0000	1.0000	1.0000	1.0000	1.0000	1.0000	1.0000	1.0000	0.9997	30
31	1.0000	1.0000	1.0000	1.0000	1.0000	1.0000	1.0000	1.0000	1.0000	1.0000	1.0000	1.0000	1.0000	1.0000	1.0000	1.0000	1.0000	0.9999	31
32	1.0000	1.0000	1.0000	1.0000	1.0000	1.0000	1.0000	1.0000	1.0000	1.0000	1.0000	1.0000	1.0000	1.0000	1.0000	1.0000	1.0000	1.0000	32

TABLE 3 Cumulative Poisson Distribution Functions

These tables give the probability of a random variable, which has a Poisson distribution with mean λ, having integer values less than or equal to x.

λ	0.10	0.20	0.30	0.40	0.50	0.60	0.70	0.80	0.90	1.00	1.20	1.40	1.60	1.80	λ
x															x
0	0.9048	0.8187	0.7408	0.6703	0.6065	0.5488	0.4966	0.4493	0.4066	0.3679	0.3012	0.2466	0.2019	0.1653	0
1	0.9953	0.9825	0.9631	0.9384	0.9098	0.8781	0.8442	0.8088	0.7725	0.7358	0.6626	0.5918	0.5249	0.4628	1
2	0.9998	0.9989	0.9964	0.9921	0.9856	0.9769	0.9659	0.9526	0.9371	0.9197	0.8795	0.8335	0.7834	0.7306	2
3	1.0000	0.9999	0.9997	0.9992	0.9982	0.9966	0.9942	0.9909	0.9865	0.9810	0.9662	0.9463	0.9212	0.8913	3
4	1.0000	1.0000	1.0000	0.9999	0.9998	0.9996	0.9992	0.9986	0.9977	0.9963	0.9923	0.9857	0.9763	0.9636	4
5	1.0000	1.0000	1.0000	1.0000	1.0000	1.0000	0.9999	0.9998	0.9997	0.9994	0.9985	0.9968	0.9940	0.9896	5
6	1.0000	1.0000	1.0000	1.0000	1.0000	1.0000	1.0000	1.0000	1.0000	0.9999	0.9997	0.9994	0.9987	0.9974	6
7	1.0000	1.0000	1.0000	1.0000	1.0000	1.0000	1.0000	1.0000	1.0000	1.0000	1.0000	0.9999	0.9997	0.9994	7
8	1.0000	1.0000	1.0000	1.0000	1.0000	1.0000	1.0000	1.0000	1.0000	1.0000	1.0000	1.0000	1.0000	0.9999	8
9	1.0000	1.0000	1.0000	1.0000	1.0000	1.0000	1.0000	1.0000	1.0000	1.0000	1.0000	1.0000	1.0000	1.0000	9

λ	2.00	2.20	2.40	2.60	2.80	3.00	3.20	3.40	3.60	3.80	4.00	4.50	5.00	5.50	λ
x															x
0	0.1353	0.1108	0.0907	0.0743	0.0608	0.0498	0.0408	0.0334	0.0273	0.0224	0.0183	0.0111	0.0067	0.0041	0
1	0.4060	0.3546	0.3084	0.2674	0.2311	0.1991	0.1712	0.1468	0.1257	0.1074	0.0916	0.0611	0.0404	0.0266	1
2	0.6767	0.6227	0.5697	0.5184	0.4695	0.4232	0.3799	0.3397	0.3027	0.2689	0.2381	0.1736	0.1247	0.0884	2
3	0.8571	0.8194	0.7787	0.7360	0.6919	0.6472	0.6025	0.5584	0.5152	0.4735	0.4335	0.3423	0.2650	0.2017	3
4	0.9473	0.9275	0.9041	0.8774	0.8477	0.8153	0.7806	0.7442	0.7064	0.6678	0.6288	0.5321	0.4405	0.3575	4
5	0.9834	0.9751	0.9643	0.9510	0.9349	0.9161	0.8946	0.8705	0.8441	0.8156	0.7851	0.7029	0.6160	0.5289	5
6	0.9955	0.9925	0.9884	0.9828	0.9756	0.9665	0.9554	0.9421	0.9267	0.9091	0.8893	0.8311	0.7622	0.6860	6
7	0.9989	0.9980	0.9967	0.9947	0.9919	0.9881	0.9832	0.9769	0.9692	0.9599	0.9489	0.9134	0.8666	0.8095	7
8	0.9998	0.9995	0.9991	0.9985	0.9976	0.9962	0.9943	0.9917	0.9883	0.9840	0.9786	0.9597	0.9319	0.8944	8
9	1.0000	0.9999	0.9998	0.9996	0.9993	0.9989	0.9982	0.9973	0.9960	0.9942	0.9919	0.9829	0.9682	0.9462	9
10	1.0000	1.0000	1.0000	0.9999	0.9998	0.9997	0.9995	0.9992	0.9987	0.9981	0.9972	0.9933	0.9863	0.9747	10
11	1.0000	1.0000	1.0000	1.0000	1.0000	0.9999	0.9999	0.9998	0.9996	0.9994	0.9991	0.9976	0.9945	0.9890	11
12	1.0000	1.0000	1.0000	1.0000	1.0000	1.0000	1.0000	0.9999	0.9999	0.9998	0.9997	0.9992	0.9980	0.9955	12
13	1.0000	1.0000	1.0000	1.0000	1.0000	1.0000	1.0000	1.0000	1.0000	1.0000	0.9999	0.9997	0.9993	0.9983	13
14	1.0000	1.0000	1.0000	1.0000	1.0000	1.0000	1.0000	1.0000	1.0000	1.0000	1.0000	0.9999	0.9998	0.9994	14
15	1.0000	1.0000	1.0000	1.0000	1.0000	1.0000	1.0000	1.0000	1.0000	1.0000	1.0000	1.0000	0.9999	0.9998	15
16	1.0000	1.0000	1.0000	1.0000	1.0000	1.0000	1.0000	1.0000	1.0000	1.0000	1.0000	1.0000	1.0000	0.9999	16
17	1.0000	1.0000	1.0000	1.0000	1.0000	1.0000	1.0000	1.0000	1.0000	1.0000	1.0000	1.0000	1.0000	1.0000	17
18	1.0000	1.0000	1.0000	1.0000	1.0000	1.0000	1.0000	1.0000	1.0000	1.0000	1.0000	1.0000	1.0000	1.0000	18
19	1.0000	1.0000	1.0000	1.0000	1.0000	1.0000	1.0000	1.0000	1.0000	1.0000	1.0000	1.0000	1.0000	1.0000	19
20	1.0000	1.0000	1.0000	1.0000	1.0000	1.0000	1.0000	1.0000	1.0000	1.0000	1.0000	1.0000	1.0000	1.0000	20

λ / x	6.0	6.5	7.0	7.5	8.0	8.5	9.0	9.5	10.0	11.0	12.0	13.0	14.0	15.0	λ / x
0	0.0025	0.0015	0.0009	0.0006	0.0003	0.0002	0.0001	0.0001	0.0000	0.0000	0.0000	0.0000	0.0000	0.0000	**0**
1	0.0174	0.0113	0.0073	0.0047	0.0030	0.0019	0.0012	0.0008	0.0005	0.0002	0.0001	0.0000	0.0000	0.0000	**1**
2	0.0620	0.0430	0.0296	0.0203	0.0138	0.0093	0.0062	0.0042	0.0028	0.0012	0.0005	0.0002	0.0001	0.0000	**2**
3	0.1512	0.1118	0.0818	0.0591	0.0424	0.0301	0.0212	0.0149	0.0103	0.0049	0.0023	0.0011	0.0005	0.0002	**3**
4	0.2851	0.2237	0.1730	0.1321	0.0996	0.0744	0.0550	0.0403	0.0293	0.0151	0.0076	0.0037	0.0018	0.0009	**4**
5	0.4457	0.3690	0.3007	0.2414	0.1912	0.1496	0.1157	0.0885	0.0671	0.0375	0.0203	0.0107	0.0055	0.0028	**5**
6	0.6063	0.5265	0.4497	0.3782	0.3134	0.2562	0.2068	0.1649	0.1301	0.0786	0.0458	0.0259	0.0142	0.0076	**6**
7	0.7440	0.6728	0.5987	0.5246	0.4530	0.3856	0.3239	0.2687	0.2202	0.1432	0.0895	0.0540	0.0316	0.0180	**7**
8	0.8472	0.7916	0.7291	0.6620	0.5925	0.5231	0.4557	0.3918	0.3328	0.2320	0.1550	0.0998	0.0621	0.0374	**8**
9	0.9161	0.8774	0.8305	0.7764	0.7166	0.6530	0.5874	0.5218	0.4579	0.3405	0.2424	0.1658	0.1094	0.0699	**9**
10	0.9574	0.9332	0.9015	0.8622	0.8159	0.7634	0.7060	0.6453	0.5830	0.4599	0.3472	0.2517	0.1757	0.1185	**10**
11	0.9799	0.9661	0.9467	0.9208	0.8881	0.8487	0.8030	0.7520	0.6968	0.5793	0.4616	0.3532	0.2600	0.1848	**11**
12	0.9912	0.9840	0.9730	0.9573	0.9362	0.9091	0.8758	0.8364	0.7916	0.6887	0.5760	0.4631	0.3585	0.2676	**12**
13	0.9964	0.9929	0.9872	0.9784	0.9658	0.9486	0.9261	0.8981	0.8645	0.7813	0.6815	0.5730	0.4644	0.3632	**13**
14	0.9986	0.9970	0.9943	0.9897	0.9827	0.9726	0.9585	0.9400	0.9165	0.8540	0.7720	0.6751	0.5704	0.4657	**14**
15	0.9995	0.9988	0.9976	0.9954	0.9918	0.9862	0.9780	0.9665	0.9513	0.9074	0.8444	0.7636	0.6694	0.5681	**15**
16	0.9998	0.9996	0.9990	0.9980	0.9963	0.9934	0.9889	0.9823	0.9730	0.9441	0.8987	0.8355	0.7559	0.6641	**16**
17	0.9999	0.9998	0.9996	0.9992	0.9984	0.9970	0.9947	0.9911	0.9857	0.9678	0.9370	0.8905	0.8272	0.7489	**17**
18	1.0000	0.9999	0.9999	0.9997	0.9993	0.9987	0.9976	0.9957	0.9928	0.9823	0.9626	0.9302	0.8826	0.8195	**18**
19	1.0000	1.0000	1.0000	0.9999	0.9997	0.9995	0.9989	0.9980	0.9965	0.9907	0.9787	0.9573	0.9235	0.8752	**19**
20	1.0000	1.0000	1.0000	1.0000	0.9999	0.9998	0.9996	0.9991	0.9984	0.9953	0.9884	0.9750	0.9521	0.9170	**20**
21	1.0000	1.0000	1.0000	1.0000	1.0000	0.9999	0.9998	0.9996	0.9993	0.9977	0.9939	0.9859	0.9712	0.9469	**21**
22	1.0000	1.0000	1.0000	1.0000	1.0000	1.0000	0.9999	0.9999	0.9997	0.9990	0.9970	0.9924	0.9833	0.9673	**22**
23	1.0000	1.0000	1.0000	1.0000	1.0000	1.0000	1.0000	0.9999	0.9999	0.9995	0.9985	0.9960	0.9907	0.9805	**23**
24	1.0000	1.0000	1.0000	1.0000	1.0000	1.0000	1.0000	1.0000	1.0000	0.9998	0.9993	0.9980	0.9950	0.9888	**24**
25	1.0000	1.0000	1.0000	1.0000	1.0000	1.0000	1.0000	1.0000	1.0000	0.9999	0.9997	0.9990	0.9974	0.9938	**25**
26	1.0000	1.0000	1.0000	1.0000	1.0000	1.0000	1.0000	1.0000	1.0000	1.0000	0.9999	0.9995	0.9987	0.9967	**26**
27	1.0000	1.0000	1.0000	1.0000	1.0000	1.0000	1.0000	1.0000	1.0000	1.0000	0.9999	0.9998	0.9994	0.9983	**27**
28	1.0000	1.0000	1.0000	1.0000	1.0000	1.0000	1.0000	1.0000	1.0000	1.0000	1.0000	0.9999	0.9997	0.9991	**28**
29	1.0000	1.0000	1.0000	1.0000	1.0000	1.0000	1.0000	1.0000	1.0000	1.0000	1.0000	1.0000	0.9999	0.9996	**29**
30	1.0000	1.0000	1.0000	1.0000	1.0000	1.0000	1.0000	1.0000	1.0000	1.0000	1.0000	1.0000	0.9999	0.9998	**30**
31	1.0000	1.0000	1.0000	1.0000	1.0000	1.0000	1.0000	1.0000	1.0000	1.0000	1.0000	1.0000	1.0000	0.9999	**31**
32	1.0000	1.0000	1.0000	1.0000	1.0000	1.0000	1.0000	1.0000	1.0000	1.0000	1.0000	1.0000	1.0000	1.0000	**32**

TABLE 4 Normal Distribution Function

The table gives the probability p that a normally distributed random variable Z, with mean = 0 and variance = 1, is less than or equal to z.

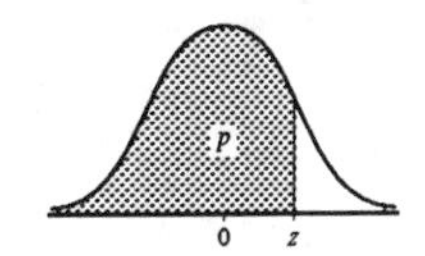

z	0.00	0.01	0.02	0.03	0.04	0.05	0.06	0.07	0.08	0.09	
0	0.50000	0.50399	0.50798	0.51197	0.51595	0.51994	0.52392	0.52790	0.53188	0.53586	0.0
0.1	0.53983	0.54380	0.54776	0.55172	0.55567	0.55962	0.56356	0.56749	0.57142	0.57535	0.1
0.2	0.57926	0.58317	0.58706	0.59095	0.59483	0.59871	0.60257	0.60642	0.61026	0.61409	0.2
0.3	0.61791	0.62172	0.62552	0.62930	0.63307	0.63683	0.64058	0.64431	0.64803	0.65173	0.3
0.4	0.65542	0.65910	0.66276	0.66640	0.67003	0.67364	0.67724	0.68082	0.68439	0.68793	0.4
0.5	0.69146	0.69497	0.69847	0.70194	0.70540	0.70884	0.71226	0.71566	0.71904	0.72240	0.5
0.6	0.72575	0.72907	0.73237	0.73565	0.73891	0.74215	0.74537	0.74857	0.75175	0.75490	0.6
0.7	0.75804	0.76115	0.76424	0.76730	0.77035	0.77337	0.77637	0.77935	0.78230	0.78524	0.7
0.8	0.78814	0.79103	0.79389	0.79673	0.79955	0.80234	0.80511	0.80785	0.81057	0.81327	0.8
0.9	0.81594	0.81859	0.82121	0.82381	0.82639	0.82894	0.83147	0.83398	0.83646	0.83891	0.9
1	0.84134	0.84375	0.84614	0.84849	0.85083	0.85314	0.85543	0.85769	0.85993	0.86214	1.0
1.1	0.86433	0.86650	0.86864	0.87076	0.87286	0.87493	0.87698	0.87900	0.88100	0.88298	1.1
1.2	0.88493	0.88686	0.88877	0.89065	0.89251	0.89435	0.89617	0.89796	0.89973	0.90147	1.2
1.3	0.90320	0.90490	0.90658	0.90824	0.90988	0.91149	0.91309	0.91466	0.91621	0.91774	1.3
1.4	0.91924	0.92073	0.92220	0.92364	0.92507	0.92647	0.92785	0.92922	0.93056	0.93189	1.4
1.5	0.93319	0.93448	0.93574	0.93699	0.93822	0.93943	0.94062	0.94179	0.94295	0.94408	1.5
1.6	0.94520	0.94630	0.94738	0.94845	0.94950	0.95053	0.95154	0.95254	0.95352	0.95449	1.6
1.7	0.95543	0.95637	0.95728	0.95818	0.95907	0.95994	0.96080	0.96164	0.96246	0.96327	1.7
1.8	0.96407	0.96485	0.96562	0.96638	0.96712	0.96784	0.96856	0.96926	0.96995	0.97062	1.8
1.9	0.97128	0.97193	0.97257	0.97320	0.97381	0.97441	0.97500	0.97558	0.97615	0.97670	1.9
2	0.97725	0.97778	0.97831	0.97882	0.97932	0.97982	0.98030	0.98077	0.98124	0.98169	2.0
2.1	0.98214	0.98257	0.98300	0.98341	0.98382	0.98422	0.98461	0.98500	0.98537	0.98574	2.1
2.2	0.98610	0.98645	0.98679	0.98713	0.98745	0.98778	0.98809	0.98840	0.98870	0.98899	2.2
2.3	0.98928	0.98956	0.98983	0.99010	0.99036	0.99061	0.99086	0.99111	0.99134	0.99158	2.3
2.4	0.99180	0.99202	0.99224	0.99245	0.99266	0.99286	0.99305	0.99324	0.99343	0.99361	2.4
2.5	0.99379	0.99396	0.99413	0.99430	0.99446	0.99461	0.99477	0.99492	0.99506	0.99520	2.5
2.6	0.99534	0.99547	0.99560	0.99573	0.99585	0.99598	0.99609	0.99621	0.99632	0.99643	2.6
2.7	0.99653	0.99664	0.99674	0.99683	0.99693	0.99702	0.99711	0.99720	0.99728	0.99736	2.7
2.8	0.99744	0.99752	0.99760	0.99767	0.99774	0.99781	0.99788	0.99795	0.99801	0.99807	2.8
2.9	0.99813	0.99819	0.99825	0.99831	0.99836	0.99841	0.99846	0.99851	0.99856	0.99861	2.9
3	0.99865	0.99869	0.99874	0.99878	0.99882	0.99886	0.99889	0.99893	0.99896	0.99900	3.0
3.1	0.99903	0.99906	0.99910	0.99913	0.99916	0.99918	0.99921	0.99924	0.99926	0.99929	3.1
3.2	0.99931	0.99934	0.99936	0.99938	0.99940	0.99942	0.99944	0.99946	0.99948	0.99950	3.2
3.3	0.99952	0.99953	0.99955	0.99957	0.99958	0.99960	0.99961	0.99962	0.99964	0.99965	3.3
3.4	0.99966	0.99968	0.99969	0.99970	0.99971	0.99972	0.99973	0.99974	0.99975	0.99976	3.4
3.5	0.99977	0.99978	0.99978	0.99979	0.99980	0.99981	0.99981	0.99982	0.99983	0.99983	3.5
3.6	0.99984	0.99985	0.99985	0.99986	0.99986	0.99987	0.99987	0.99988	0.99988	0.99989	3.6
3.7	0.99989	0.99990	0.99990	0.99990	0.99991	0.99991	0.99992	0.99992	0.99992	0.99992	3.7
3.8	0.99993	0.99993	0.99993	0.99994	0.99994	0.99994	0.99994	0.99995	0.99995	0.99995	3.8
3.9	0.99995	0.99995	0.99996	0.99996	0.99996	0.99996	0.99996	0.99996	0.99997	0.99997	3.9

TABLE 5 Percentage Points of the Normal Distribution

The table gives the values of z satisfying $P(Z \le z) = p$, where Z is the normally distributed random variable with mean = 0 and variance = 1.

p	0.00	0.01	0.02	0.03	0.04	0.05	0.06	0.07	0.08	0.09	
0.5	0.0000	0.0251	0.0502	0.0753	0.1004	0.1257	0.1510	0.1764	0.2019	0.2275	**0.5**
0.6	0.2533	0.2793	0.3055	0.3319	0.3585	0.3853	0.4125	0.4399	0.4677	0.4958	**0.6**
0.7	0.5244	0.5534	0.5828	0.6128	0.6433	0.6745	0.7063	0.7388	0.7722	0.8064	**0.7**
0.8	0.8416	0.8779	0.9154	0.9542	0.9945	1.0364	1.0803	1.1264	1.1750	1.2265	**0.8**
0.9	1.2816	1.3408	1.4051	1.4758	1.5548	1.6449	1.7507	1.8808	2.0537	2.3263	**0.9**
p	**0.000**	**0.001**	**0.002**	**0.003**	**0.004**	**0.005**	**0.006**	**0.007**	**0.008**	**0.009**	
0.95	1.6449	1.6546	1.6646	1.6747	1.6849	1.6954	1.7060	1.7169	1.7279	1.7392	**0.95**
0.96	1.7507	1.7624	1.7744	1.7866	1.7991	1.8119	1.8250	1.8384	1.8522	1.8663	**0.96**
0.97	1.8808	1.8957	1.9110	1.9268	1.9431	1.9600	1.9774	1.9954	2.0141	2.0335	**0.97**
0.98	2.0537	2.0749	2.0969	2.1201	2.1444	2.1701	2.1973	2.2262	2.2571	2.2904	**0.98**
0.99	2.3263	2.3656	2.4089	2.4573	2.5121	2.5758	2.6521	2.7478	2.8782	3.0902	**0.99**

TABLE 6 χ^2 Distribution Function

This table gives the values of x satisfying $P(X \leq x) = p$ where X is a χ^2 random variable with υ degrees of freedom.

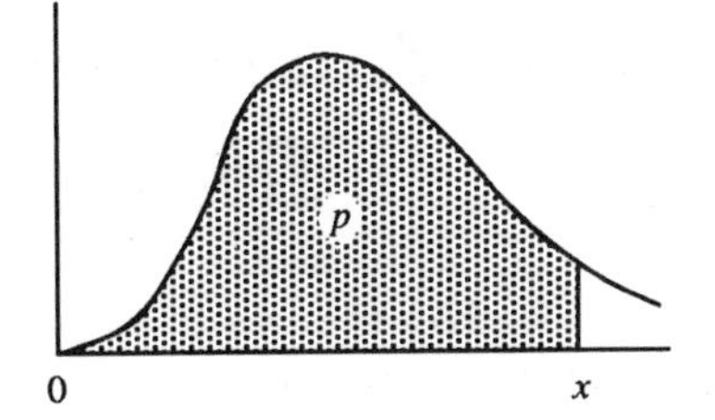

p / υ	0.005	0.010	0.025	0.050	0.100	0.900	0.950	0.975	0.990	0.995	p / υ
1	0.00004	0.0002	0.001	0.004	0.016	2.706	3.841	5.024	6.635	7.879	1
2	0.010	0.020	0.051	0.103	0.211	4.605	5.991	7.378	9.210	10.597	2
3	0.072	0.115	0.216	0.352	0.584	6.251	7.815	9.348	11.345	12.838	3
4	0.207	0.297	0.484	0.711	1.064	7.779	9.488	11.143	13.277	14.860	4
5	0.412	0.554	0.831	1.145	1.610	9.236	11.070	12.833	15.086	16.750	5
6	0.676	0.872	1.237	1.635	2.204	10.645	12.592	14.449	16.812	18.548	6
7	0.989	1.239	1.690	2.167	2.833	12.017	14.067	16.013	18.475	20.278	7
8	1.344	1.646	2.180	2.733	3.490	13.362	15.507	17.535	20.090	21.955	8
9	1.735	2.088	2.700	3.325	4.168	14.684	16.919	19.023	21.666	23.589	9
10	2.156	2.558	3.247	3.940	4.865	15.987	18.307	20.483	23.209	25.188	10
11	2.603	3.053	3.816	4.575	5.578	17.275	19.675	21.920	24.725	26.757	11
12	3.074	3.571	4.404	5.226	6.304	18.549	21.026	23.337	26.217	28.300	12

TABLE 7 Critical Values of the Product Moment Correlation Coefficient

This table gives the critical values, for different significance levels, of the sample product moment correlation coefficient, r, for varying sample size, n.

One tail	10%	5%	2.5%	1%	0.5%	
Two tail	20%	10%	5%	2%	1%	
n						
4	0.8000	0.9000	0.9500	0.9800	0.9900	4
5	0.6870	0.8054	0.8783	0.9343	0.9587	5
6	0.6084	0.7293	0.8114	0.8822	0.9172	6
7	0.5509	0.6694	0.7545	0.8329	0.8745	7
8	0.5067	0.6215	0.7067	0.7887	0.8343	8
9	0.4716	0.5822	0.6664	0.7498	0.7977	9
10	0.4428	0.5494	0.6319	0.7155	0.7646	10
11	0.4187	0.5214	0.6021	0.6851	0.7348	11
12	0.3981	0.4973	0.5760	0.6581	0.7079	12
13	0.3802	0.4762	0.5529	0.6339	0.6835	13
14	0.3646	0.4575	0.5324	0.6120	0.6614	14
15	0.3507	0.4409	0.5140	0.5923	0.6411	15
16	0.3383	0.4259	0.4973	0.5742	0.6226	16
17	0.3271	0.4124	0.4821	0.5577	0.6055	17
18	0.3170	0.4000	0.4683	0.5425	0.5897	18
19	0.3077	0.3887	0.4555	0.5285	0.5751	19
20	0.2992	0.3783	0.4438	0.5155	0.5614	20

TABLE 8 Critical Values of Spearman's Rank Correlation Coefficient

This table gives the critical values, for different significance levels, of Spearman's rank correlation coefficient, r_s, for varying sample sizes, n.

One tail Two tail	10% 20%	5% 10%	2.5% 5%	1% 2%	0.5% 1%	
n						
4	1.0000	1.0000	1.0000	1.0000	1.0000	**4**
5	0.7000	0.9000	0.9000	1.0000	1.0000	**5**
6	0.6571	0.7714	0.8286	0.9429	0.9429	**6**
7	0.5714	0.6786	0.7857	0.8571	0.8929	**7**
8	0.5476	0.6429	0.7381	0.8095	0.8571	**8**
9	0.4833	0.6000	0.6833	0.7667	0.8167	**9**
10	0.4424	0.5636	0.6485	0.7333	0.7818	**10**
11	0.4182	0.5273	0.6091	0.7000	0.7545	**11**
12	0.3986	0.5035	0.5874	0.6713	0.7273	**12**
13	0.3791	0.4780	0.5604	0.6484	0.6978	**13**
14	0.3670	0.4593	0.5385	0.6220	0.6747	**14**
15	0.3500	0.4429	0.5179	0.6000	0.6536	**15**
16	0.3382	0.4265	0.5029	0.5824	0.6324	**16**
17	0.3271	0.4124	0.4821	0.5577	0.6055	**17**
18	0.3170	0.4000	0.4683	0.5425	0.5897	**18**
19	0.3077	0.3887	0.4555	0.5285	0.5751	**19**
20	0.2992	0.3783	0.4438	0.5155	0.5614	**20**

SOURCES OF DATA

Statistical data can be gathered from the following official bodies:

Central Statistics Office 071 270 6363.

The range of publications is described in Section 2.1 and all their publications are available from your local HMSO supplier.

The address is as follows

CSO
Cabinet Office
Great George Street
London SW1

Department of Employment 071 273 6969

The Department of Employment publishes the *Employment Gazette* monthly, giving details of employment, unemployment, regional variations, earnings, retail price index and redundancy. The *Gazette* is available from

Harrington Kilbride (0908 371981)
1st Floor
Stephenson House
Brunel Centre
Bletchley
Milton Keynes MK2 2EW

Office of Population Censuses and Surveys 071 243 0262

The Office collects and publishes census data; their publications are available from your local HMSO supplier.

The address is as follows:

Office of Population Censuses and Surveys
St Catherine's House
10 Kingsway
London WC2.

FURTHER READING

There have been many books written for A level syllabuses in Statistics in recent years, and you will find any of the following texts useful for extra worked examples and exercises on topics in the Statistics Module in the AEB A/AS level Mathematics framework.

1. **A Concise Course in A-Level Statistics**, 2nd edition, *J. Crawshaw* and *J.Chambers* , 1990 (Stanley Thornes) 0 85950 378 X

2. **A Basic Course in Statistics**, 3rd edition, *G. Clarke* and *D. Cooke*, 1992 (Hodder Headline) 0 34056 772 4

3. **Statistics for Advanced Level**, 2nd edition, *J. Miller*, 1989 (Cambridge) 0 521 36772 7

4. **Statistics**, 2nd edition, *F. Owen* and *R. Jones*, 1982 (Longman) 0 273 02639 9

5. **Probability**, *D. Rountree*, 1984 (Edward Arnold) 0 7131 2787 2

6. **Revision Statistics**, *M. Godfrey, E. Roebuck* and *A. Sherlock*, 1986 (Edward Arnold) 0 7131 3540 9

7. **Statistics for Advanced Level Mathematics**, *I. Evans*, 1984 (Hodder and Stoughton) 0 340 32764 2

8. **Advanced Level Statistics**, 2nd edition, *A. Francis*, 1988 (Stanley Thornes) 0 85950 813 7

INDEX

Alternative hypothesis 190
Approximations
  normal to binomial and Poisson 162
  Poisson to binomial 126
Arithmetic mean 65

Bar charts 51
Bias 35
Binomial distribution 99
  mean and variance 107
Bivariate distributions 234
Box and whisker plot 72
Buffon's needle 24

Census 35
Central Limit Theorem 169, 178
Chi-squared distribution 203
  degrees of freedom 205
  significance testing 206
Cluster sample 41
Combined events 9
Complement 9
Composite bar chart 52
Conditional probability 16
Confidence intervals 181
Contingency tables 208
Continuity correction 137
Continuous data 31, 87
Continuous probability distributions 131
  mean and variance 141
  mode, median and quartiles 143
Correlation 215
Covariance 217
Critical region 197
Cumulative distribution function 139
Cumulative frequency curve 63

Data
  collection 29
  continuous 31, 87
  discrete 31, 87
  grouped 62
  primary and secondary 31
  qualitative 31
  quantitive 31
  range 69
  sources 33
  spread 68, 73

Degrees of freedom 205
Dependence 21
Discrete data 31, 87
Discrete probability distributions 87
Distributions
  binomial 99
  chi-squared 203
  normal 151
  Poisson 115
  probability function 131
  rectangular 146
  sample mean $\bar{x}$ 177
  uniform 95

Empirical probability 5, 7
Estimation 173
Events
  dependent 21
  exhaustive 12
  independent 20
  mutually exclusive 10
Exhaustive probabilities 12
Expectation 89
Extrapolating 232

Frequency
  curve 55
  polygon 55
  relative 9
  table 48

Grouped data 62

Histogram 53
Hypothesis testing 189
  critical region 197
  one tailed test 196
  two tailed test 197

Independence 20
Inter-quartile range 69
Interpolating 232
Intersection 9

Least squares method 74, 231
Line graphs 57
Linear interpolation 63
Linear regression 230
Lower quartile 69

Mean 65
- arithmetic 65
- binomial 107
- confidence interval 181
- continuous probability distributions 141
- hypothesis testing 195
- Poisson 125
- random variable 89
- rectangular/uniform distribution 97, 146

Median 61, 144
Mode 60, 143
Mutually exclusive events 10, 12

Normal distribution 151
- approximation to other distributions 162
- using tables 155

Null hypothesis 190

Parameter 175
Pascal's triangle 102
Pie charts 56
Poisson distributions 115, 118
- approximation to binomial 126
- combining variables 121
- mean and variance 125

Population parameters 175
Primary data 31
Probability
- combined events 9
- conditional 16
- density function 92, 155
- distribution 92
- distribution function 131
- empirical 5, 6, 7
- exhaustive 12
- experiment 5
- observation 7
- symmetry 1
- theoretical 1

Product moment correlation coefficient 219
Purposive sample 41

Qualitative data 31
Quantitative data 31
Quartiles 69, 144

Random numbers 42
Random sampling 41, 174
Random variable 87
Range 69
Rectangular distribution 146
Regression 230
Relative frequency 9
Residuals 231

Sample space 3
Sampling 35
- random 41, 174
- systematic 41
- stratified 41
- purposive 41
- cluster 41
- means 175

Scattergram 57, 216
Secondary data 31
Semi inter-quartile range 69
Sign test 192
Significance testing 206, 221, 227
Spearman's rank correlation coefficient 224
Spread 68, 73
Standard deviation 73
Standardised normal distribution 156
Statistic 175
Stem and leaf diagrams 50
Stratified sample 41
Survey 36
Symmetry 1
Systematic sampling 41

Tied ranks 228
Tree diagrams 13

Unbiased estimator 175
Uniform distribution 95
Union 10
Upper class boundary 63
Upper quartile 69

Variable, random 87
Variance 90
- binomial 107
- continuous probability distribution 141
- Poisson 125
- rectangular/uniform distribution 97, 146

Venn diagrams 10

Yate's continuity correction 209